Triceratops

millions of years ago

| -240 | -220 | -200 | -180 | -160 | -140 | -120 | -100 |

-201.3   -145   -66

| Triassic | Jurassic | Cretaceous |

Time Period

Triceratops was a large herbivorous dinosaur that inhabited North America. Its name means "Three-Horned Face." It is supposed that the head was so heavy that Triceratops walked with its head close to the ground. The large frill on its head was sturdy and became larger as the dinosaur grew. Also, it is believed that Triceratops might have had the frill to attract its partner or mates.

● About 30 feet or 9 meters long and weighed around 5.4 tons.

After finishing each day's math and reading exercises, paste a sticker in the appropriate place below. You can also try to complete each day's fitness challenge after your math and reading exercises. Don't forget to fill out and complete the Summer Bucket List below. Stay active and learn more this summer with Kumon!

To parents: If your child seems to have difficulty pasting the stickers or writing the dates, you can offer to help. Your child may need help completing the fitness challenges, if your child needs help please demonstrate the exercise for him or her. Encourage your child to use the fitness challenges included in the progress calendar as a starting point for his or her recommended 30 minutes of daily physical activity.

When he or she has completed all of the pages, offer lots of praise and paste the biggest sticker next to "Goal!"

Finally, please fill out your child's name and sign your name in the blank.

Date / /

Math	Reading

2

Do 5 jumping jacks.

Date / /

Math	Reading

3

Do 5 toe touches.

Date / /

Math	Reading

4

Do 5 arm circles.

Date / /

Math	Reading

8

Do 10 crisscross jumps.

Date / /

Math	Reading

9

Do 5 sit-ups.

Date / /

Math	Reading

10

Do a plank for 15 seconds.

Date / /

Math	Reading

14

Do 15 jumping jacks.

Date / /

Math	Reading

18

Go for a walk.

Date / /

Math	Reading

22

Do 10 sit-ups.

Summer Bucket List

CREATE A LIST OF ACTIVITIES TO DO THIS SUMMER!

- Build a blanket fort.
  - ....
  - ....
- Play a board game.
  - ....
  - ....
- Bake cookies with a family member.
  - ....
  - ....
- Read a book out loud with your family.
  - ....
  - ....
- Design/draw your dream house or bedroom.
  - ....
  - ....

STAY ACTIVE THIS SUMMER WITH KUMON!

Date / /

Math | Reading

23

Do 5 frog jumps.

Date / /

Math | Reading

24

Ask a family member to go for a hike.

Date / /

Math | Reading

25

Do 5 push ups.

Date / /

Math | Reading

29

Go for a walk.

Date / /

Math | Reading

30

Do 10 leg raises with each leg.

Date / /

Math | Reading

31

Jump rope for 5 minutes.

Date / /

Math | Reading

33

Do butterfly kicks for 30 seconds.

Date / /

Math | Reading

34

Do a plank for 20 seconds.

Date / /

Math | Reading

35

Do 10 bicycle kicks.

Date / /

Math | Reading

37

Do 15 toe touches.

Date / /

Math | Reading

38

Go for a walk.

Date / /

Math | Reading

39

Skip for 5 minutes.

Date / /

Math | Reading

41

Do 20 jumping jacks.

Date / /

Math | Reading

42

Do 5 push ups.

Date / /

Math | Reading

43

Do 10 sit ups.

Date / /

Math | Reading

44

Do 10 crisscross feet jumps.

Date / /

Math | Reading

45

Go for a walk.

Goa is her
on completing Summer Re
Presented on _____, 20

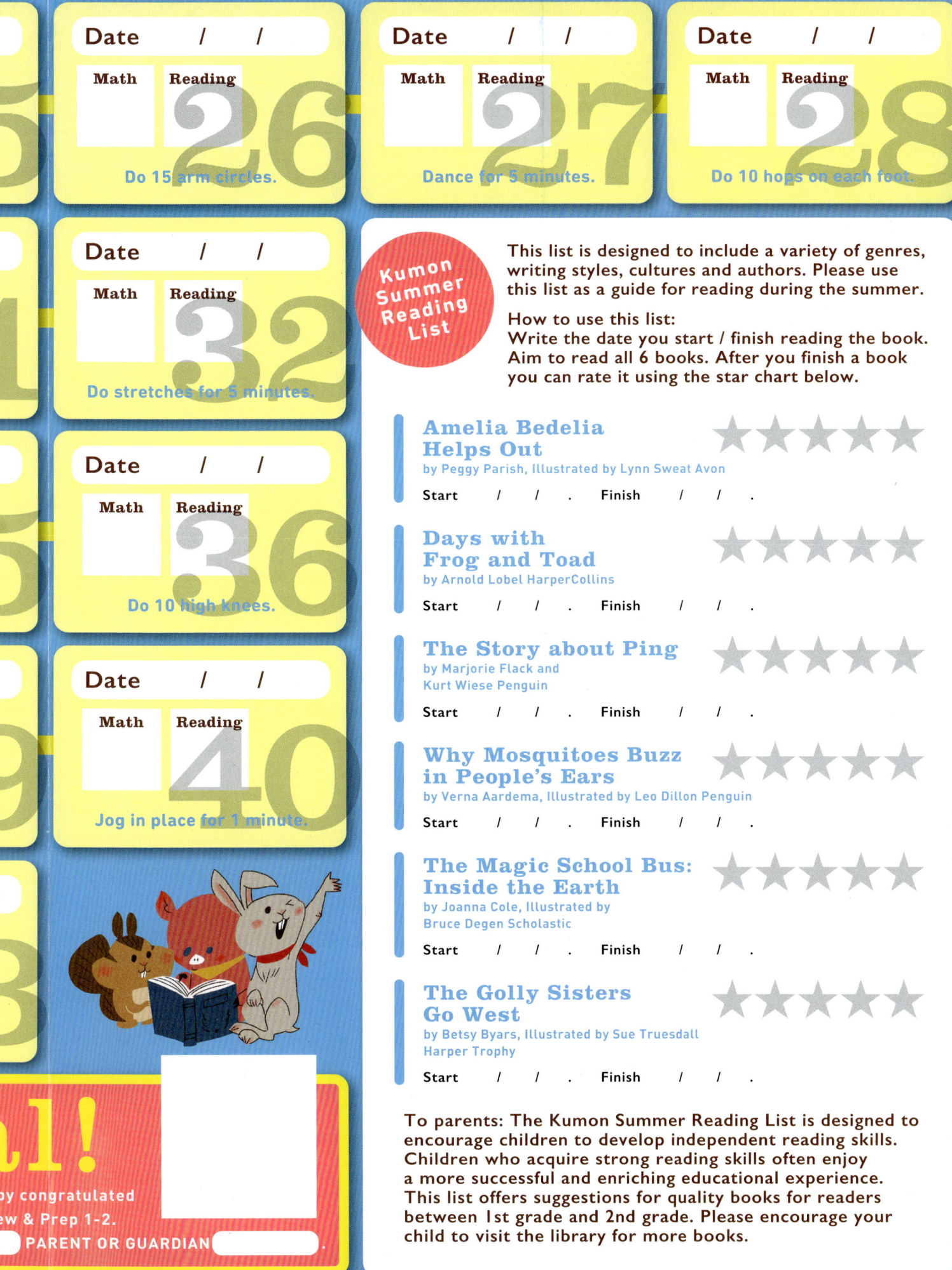

Date / /

Math   Reading

26

Do 15 arm circles.

Date / /

Math   Reading

27

Dance for 5 minutes.

Date / /

Math   Reading

28

Do 10 hops on each foot.

Date / /

Math   Reading

32

Do stretches for 5 minutes.

Date / /

Math   Reading

36

Do 10 high knees.

Date / /

Math   Reading

40

Jog in place for 1 minute.

Kumon Summer Reading List

This list is designed to include a variety of genres, writing styles, cultures and authors. Please use this list as a guide for reading during the summer.

How to use this list:
Write the date you start / finish reading the book. Aim to read all 6 books. After you finish a book you can rate it using the star chart below.

Amelia Bedelia Helps Out
by Peggy Parish, Illustrated by Lynn Sweat Avon
★★★★★
Start / / . Finish / / .

Days with Frog and Toad
by Arnold Lobel HarperCollins
★★★★★
Start / / . Finish / / .

The Story about Ping
by Marjorie Flack and Kurt Wiese Penguin
★★★★★
Start / / . Finish / / .

Why Mosquitoes Buzz in People's Ears
by Verna Aardema, Illustrated by Leo Dillon Penguin
★★★★★
Start / / . Finish / / .

The Magic School Bus: Inside the Earth
by Joanna Cole, Illustrated by Bruce Degen Scholastic
★★★★★
Start / / . Finish / / .

The Golly Sisters Go West
by Betsy Byars, Illustrated by Sue Truesdall Harper Trophy
★★★★★
Start / / . Finish / / .

To parents: The Kumon Summer Reading List is designed to encourage children to develop independent reading skills. Children who acquire strong reading skills often enjoy a more successful and enriching educational experience. This list offers suggestions for quality books for readers between 1st grade and 2nd grade. Please encourage your child to visit the library for more books.

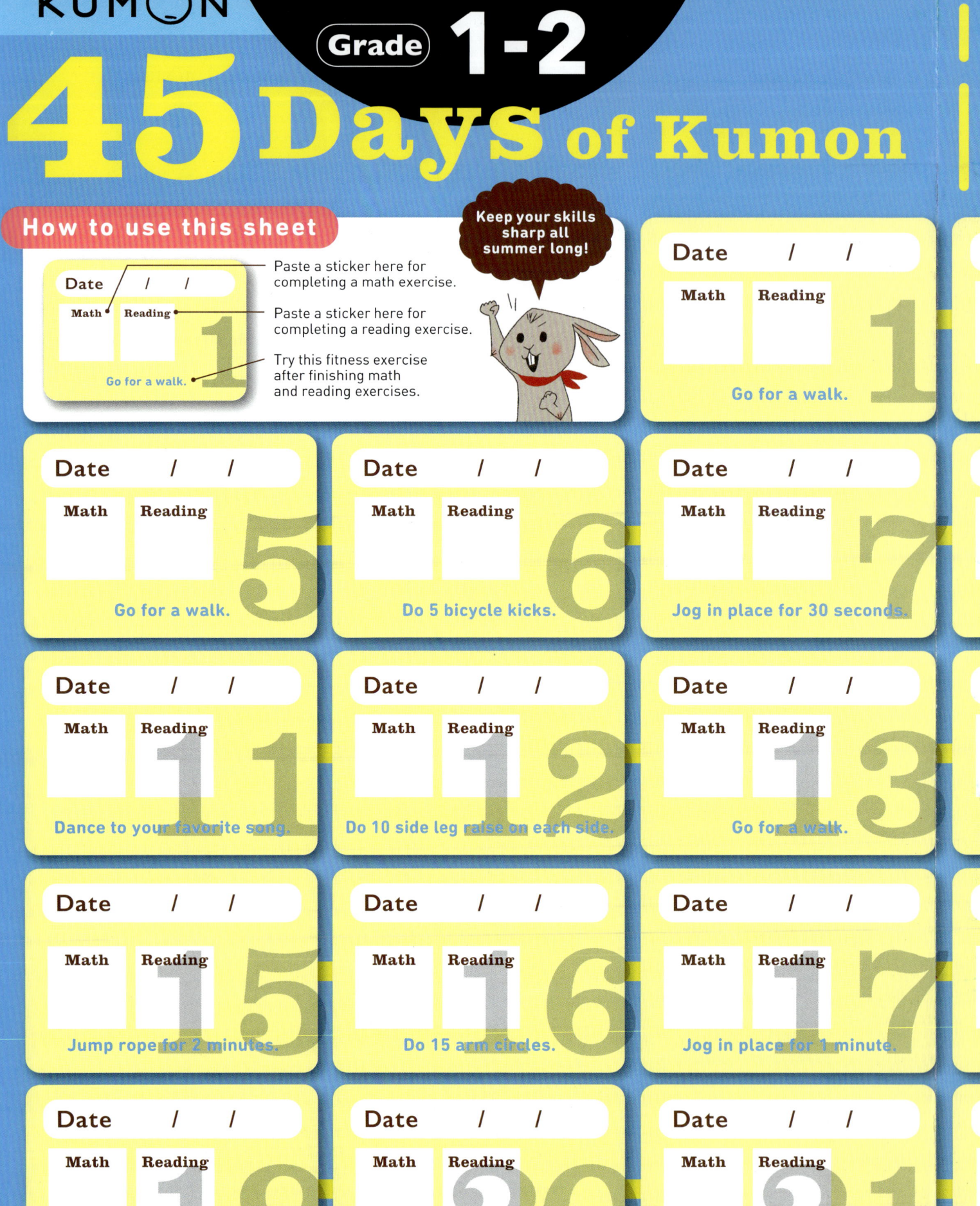

Short Vowel Sounds

"i" & "o" sounds

Date   /   /

Name

① Trace the words below. Then circle the picture that rhymes 5 points per question
with each word.

(1) sit

(2) lip

(3) pin

(4) big

(5) hog

(6) box

(7) job

(8) mop

② Complete each sentence using a word from the box. Use the 10 points per question
pictures as hints.

hit   kid   hot   log   ox   hop

(1) Bunnies like to _hop_.

(2) He _hit_ the ball!

(3) The stove is _hot_.

(4) I sat on the _log_.

(5) My little sister is a _kid_.

(6) The _ox_ pulls a cart.

Level ★

Score

100

/100

Math
DAY
4

1 Read each number sentence as you trace it or say the number sentence aloud as you add. Use the examples as hints.

5 points per question

(1) $1 + 3 = 4$
One plus three equals four

(2) $2 + 3 = 5$
Two plus three equals five

(3) $3 + 3 = 6$
Three plus three equals six

(4) $4 + 3 = 7$
Four plus three equals seven

(5) $5 + 3 = 8$
Five plus three equals eight

(6) $6 + 3 = 9$

(7) $7 + 3 = 10$

(8) $8 + 3 = 11$

(9) $9 + 3 = 12$

(10) $10 + 3 = 13$

2 Add.

5 points per question

(1) $3 + 3 = 6$

(2) $1 + 3 = 4$

(3) $2 + 3 = 5$

(4) $4 + 3 = 7$

(5) $5 + 3 = 8$

(6) $10 + 3 = 13$

(7) $9 + 3 = 12$

(8) $8 + 3 = 11$

(9) $6 + 3 = 10$

(10) $7 + 3 = 9$

Reading
DAY
3

Short Vowel Sounds
"a" & "e" vowel sounds

Date
7/22/23

Name
Margot

Level ⭐
Score
100

/100

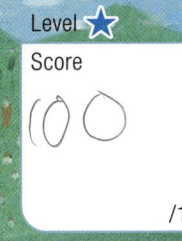

1 Trace the words below. Then circle the picture that rhymes with each word. **5** points per question

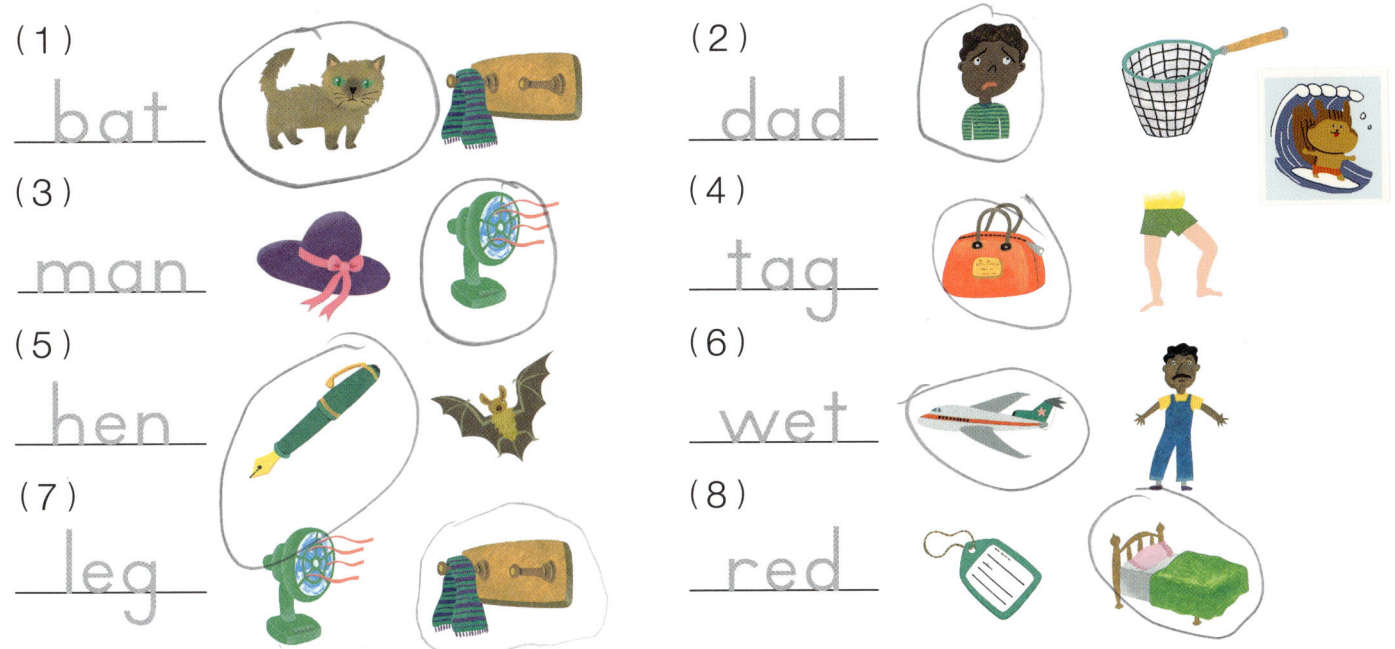

(1)
bat

(2)
dad

(3)
man

(4)
tag

(5)
hen

(6)
wet

(7)
leg

(8)
red

2 Complete each sentence using a word from the box. Use the pictures as hints. **10** points per question

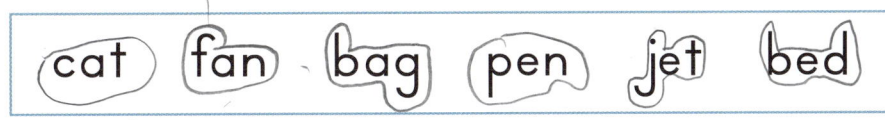

cat   fan   bag   pen   jet   bed

(1) The ___cat___ climbed the tree.

(2) I wrote my name with a _pen_.

(3) She carries a _bag_.

(4) Turn on the _fan_.

(5) The _jet_ flew over us.

(6) We jumped on the _bed_.

6

Adding 2

Level ★

Score

100

/100

Math

DAY

3

Date: 7 / 22 / 23

Name: Margot

---

1 Read each number sentence as you trace it and add. Use the examples as hints.   5 points per question

(1) $1 + 2 = 3$
One plus two equals three

(2) $2 + 2 = 4$
Two plus two equals four

(3) $3 + 2 = 5$
Three plus two equals five

(4) $4 + 2 = 6$
Four plus two equals six

(5) $5 + 2 = 7$
Five plus two equals seven

(6) $6 + 2 = 8$
Six plus two equals eight

(7) $7 + 2 = 9$
Seven plus two equals nine

(8) $8 + 2 = 10$

(9) $9 + 2 = 11$

(10) $10 + 2 = 12$

---

2 Add.   5 points per question

(1) $3 + 2 = 5$

(2) $6 + 2 = 8$

(3) $7 + 2 = 9$

(4) $1 + 2 = 3$

(5) $2 + 2 = 4$

(6) $4 + 2 = 6$

(7) $5 + 2 = 7$

(8) $10 + 2 = 12$

(9) $9 + 2 = 11$

(10) $8 + 2 = 10$

Writing Lowercase Letters
Writing a to z

Date   /   /

Name

Level ★
Score

/10

① Trace the letters a to z while saying each letter aloud.

100 points for com

  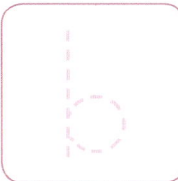  

 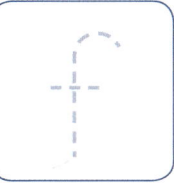  

 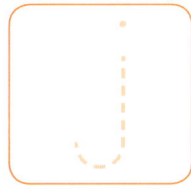  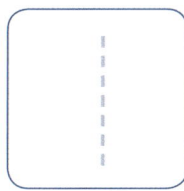

Adding 1

Level ⭐

Score

/100

Math
DAY

2

Date　　/　　/

Name

1 Read each number sentence as you trace it.　　5 points per question

(1) 1 + 1 = 2
One plus one equals two

(2) 2 + 1 = 3
Two plus one equals three

(3) 3 + 1 = 4
Three plus one equals four

(4) 4 + 1 = 5
Four plus one equals five

(5) 5 + 1 = 6
Five plus one equals six

(6) 6 + 1 = 7
Six plus one equals seven

(7) 7 + 1 = 8
Seven plus one equals eight

(8) 8 + 1 = 9
Eight plus one equals nine

(9) 9 + 1 = 10
Nine plus one equals ten

(10) 10 + 1 = 11
Ten plus one equals eleven

2 Add.　　5 points per question

(1) 3 + 1 =

(2) 6 + 1 =

(3) 7 + 1 =

(4) 2 + 1 =

(5) 1 + 1 =

(6) 4 + 1 =

(7) 5 + 1 =

(8) 10 + 1 =

(9) 9 + 1 =

(10) 8 + 1 =

You're off to a great start!

Writing Uppercase Letters
Writing A to Z

Date / /

Name

① **Trace the letters A to Z while saying each letter aloud.**   100 points for comp

   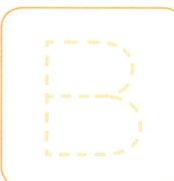    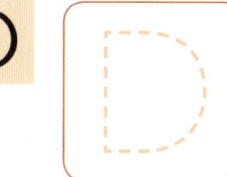

   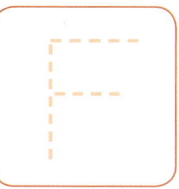 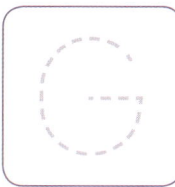  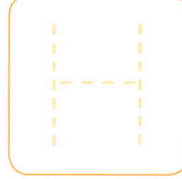

  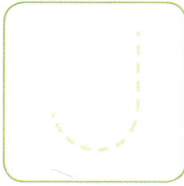   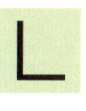 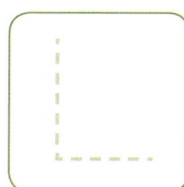

    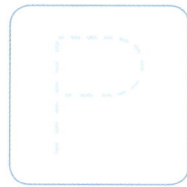

   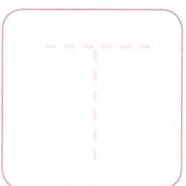

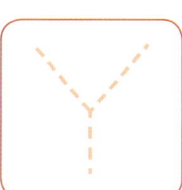 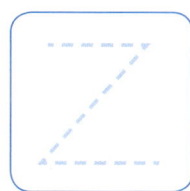

Date: 7 / 22 / 23

Name: Margot

/100

1 Fill in the missing numbers while saying each number aloud.

100 points for completion

1	2	3	4	5	6	7	8	9	10
1	2	3	4	5	6	7	8	9	10
11	12	13	14	15	16	17	18	19	20
11	12	13	14	15	16	17	18	19	20
21	22	23	24	25	26	27	28	29	30
21	22	23	24	25	26	27	28	29	30
31	32	33	34	35	36	37	38	39	40
31	32	33	34	35	36	37			
41	42	43	44	45	46	47	48	49	50
51	52	53	54	55	56	57	58	59	60
61	62	63	64	65	66	67	68	69	70
71	72	73	74	75	76	77	78	79	80
81	82	83	84	85	86	87	88	89	90
91	92	93	94	95	96	97	98	99	100
101	102	103	104	105	106	107	108	109	110
111	112	113	114	115	116	117	118	119	120

Reading
DAY
5

Short Vowel Sounds
"u" sounds & review

Level ⭐
Score

/100

Date   /   /

Name

① Trace the words below. Then circle the picture that rhymes 5 points per question
with each word.

(1) ___cub___

(2) ___cut___

(3) ___pup___

(4) ___bug___

② Complete each sentence using a word from the box. Use the 20 points per question
pictures as hints.

| tub   cup   hug |

(1) I give my mom a _____.

(2) Don't tip the _____.

(3) Fill the _____ with water.

③ Trace the consonants and write the correct vowels to finish 5 points per question
each pair of words below.

(1) p___n   p___n

(2) b___g   b___g

(3) s_a_t   s___t

(4) m___n   m___n

Adding 4

Level ⭐

Score

Math
DAY

5

Date   /   /

Name

100 /100

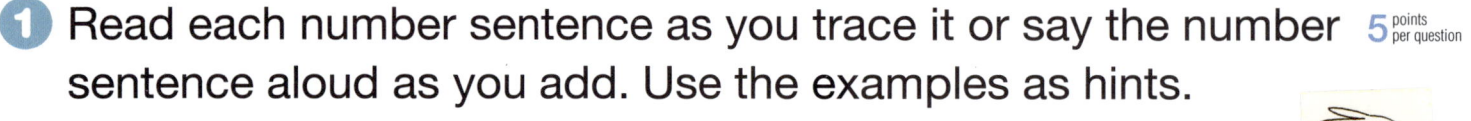

1 Read each number sentence as you trace it or say the number sentence aloud as you add. Use the examples as hints. **5** points per question

(1) $1 + 4 = 5$
One   plus   four   equals   five

(2) $2 + 4 = 6$
Two   plus   four   equals   six

(3) $3 + 4 = 7$
Three   plus   four   equals   seven

(4) $4 + 4 = 8$

(5) $5 + 4 = 9$

(6) $6 + 4 = 10$

(7) $7 + 4 = 11$

(8) $8 + 4 = 12$

(9) $9 + 4 = 13$

(10) $10 + 4 = 14$

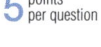

2 Add. **5** points per question

(1) $3 + 4 = 7$

(2) $1 + 4 = 5$

(3) $2 + 4 = 6$

(4) $4 + 4 = 8$

(5) $5 + 4 = 9$

(6) $10 + 4 = 14$

(7) $9 + 4 = 13$

(8) $8 + 4 = 12$

(9) $6 + 4 = 10$

(10) $7 + 4 = 11$

Date
/ /

Name

1 Add. Use the examples as hints.

3 points per question

(1) 1 + 5 = 6
(2) 2 + 5 = 7
(3) 3 + 5 =
(4) 4 + 5 =
(5) 5 + 5 =

(6) 6 + 5 =
(7) 7 + 5 =
(8) 8 + 5 =
(9) 9 + 5 =
(10) 10 + 5 =

2 Add.

3 points per question

(1) 1 + 6 =
(2) 2 + 6 =
(3) 3 + 6 =
(4) 4 + 6 =
(5) 5 + 6 =

(6) 6 + 6 =
(7) 7 + 6 =
(8) 8 + 6 =
(9) 9 + 6 =
(10) 10 + 6 =

3 Add.

4 points per question

(1) 2 + 5 =
(2) 3 + 5 =
(3) 3 + 6 =
(4) 4 + 6 =
(5) 6 + 6 =

(6) 7 + 5 =
(7) 8 + 5 =
(8) 9 + 6 =
(9) 10 + 6 =
(10) 9 + 5 =

Short Vowel Sounds
Review

Date   /   /

Name

① Write the correct vowels to finish each pair of words below. 5 points per question

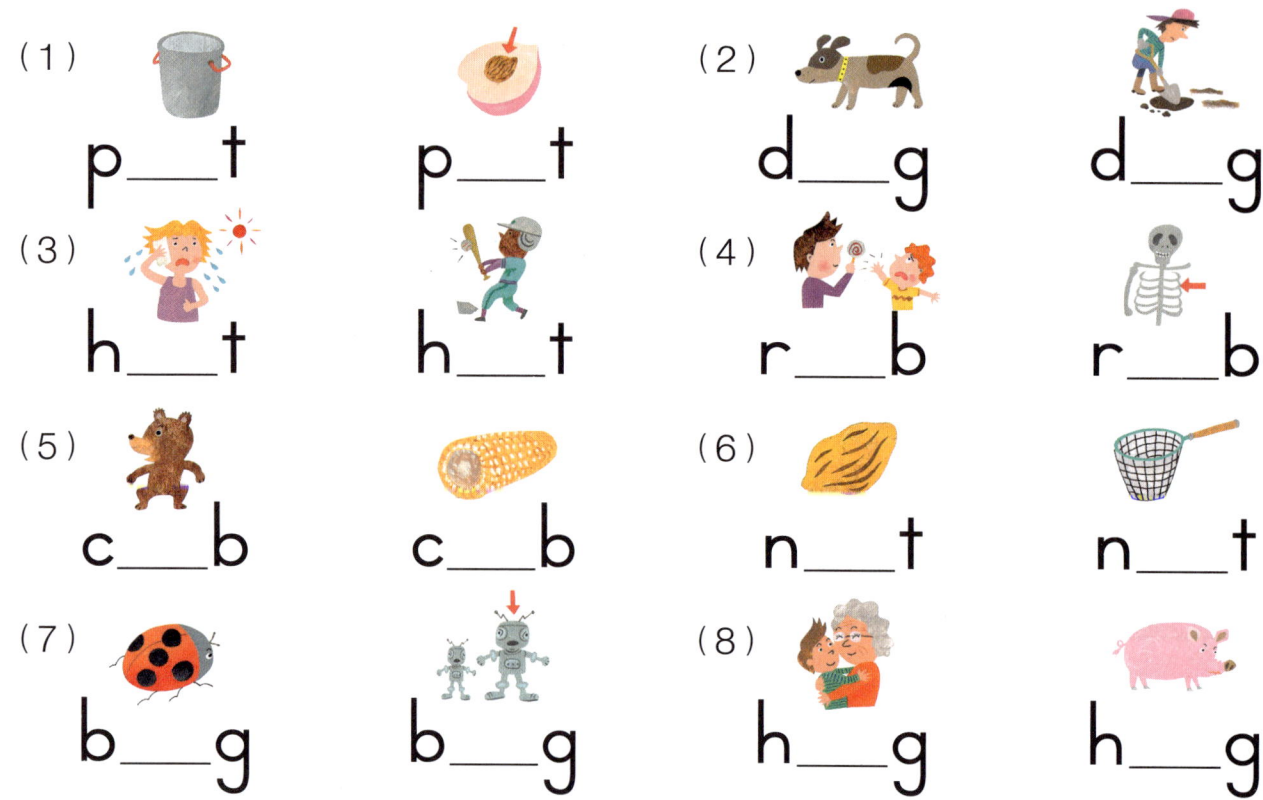

(1) p__t   p__t   (2) d__g   d__g

(3) h__t   h__t   (4) r__b   r__b

(5) c__b   c__b   (6) n__t   n__t

(7) b__g   b__g   (8) h__g   h__g

② Complete each sentence using a word from the box. Use the pictures as hints. 12 points per question

| hot dig sun pan red |

(1) The man holds a ____.

(2) My bed is ____.

(3) The pig likes to ____.

(4) The tea pot is ____.

(5) We have fun in the ____.

If you want more rhyming-words practice, check out Kumon's *My Book of RHYMING WORDS* or *My Book of RHYMING WORDS & PHRASES*.

Level ⭐

Score

/100

Math

DAY

7

Date / /

Name

1 Add. Use the examples as hints.

3 points per question

(1) $1 + 7 = 8$

(2) $2 + 7 = 9$

(3) $3 + 7 =$

(4) $4 + 7 =$

(5) $5 + 7 =$

(6) $6 + 7 =$

(7) $7 + 7 =$

(8) $8 + 7 =$

(9) $9 + 7 =$

(10) $10 + 7 =$

2 Add.

3 points per question

(1) $1 + 8 =$

(2) $2 + 8 =$

(3) $3 + 8 =$

(4) $4 + 8 =$

(5) $5 + 8 =$

(6) $6 + 8 =$

(7) $7 + 8 =$

(8) $8 + 8 =$

(9) $9 + 8 =$

(10) $10 + 8 =$

3 Add.

4 points per question

(1) $3 + 7 =$

(2) $4 + 7 =$

(3) $2 + 8 =$

(4) $3 + 8 =$

(5) $1 + 7 =$

(6) $7 + 8 =$

(7) $8 + 8 =$

(8) $9 + 7 =$

(9) $7 + 7 =$

(10) $10 + 8 =$

Digraphs
wh, sh, ch & th

Date   /   /

Name

① Trace the words below. Color the boxes of the words with the same digraph to connect three in a row like tic-tac-toe.   25 points per question

(1) words that start with "wh"

chop	sheep	thin
wheel	what	whale
three	shop	chair

(2) words that start with "sh"

whale	sheep	chin
chain	shark	think
thick	ship	wheat

(3) words that start with "ch"

shark	what	chop
thin	ship	chair
whale	think	chin

(4) words that start with "th"

wheat	shop	think
chin	thin	chair
thick	wheel	shark

Adding 9 & 10

Level ★★

Score

/100

Math
DAY
8

1 Add. Use the example as a hint.

3 points per question

(1) $1 + 9 = 10$

(2) $2 + 9 =$

(3) $3 + 9 =$

(4) $4 + 9 =$

(5) $5 + 9 =$

(6) $6 + 9 =$

(7) $7 + 9 =$

(8) $8 + 9 =$

(9) $9 + 9 =$

(10) $10 + 9 =$

2 Add. Use the example as a hint.

3 points per question

(1) $1 + 10 = 11$

(2) $2 + 10 =$

(3) $3 + 10 =$

(4) $4 + 10 =$

(5) $5 + 10 =$

(6) $6 + 10 =$

(7) $7 + 10 =$

(8) $8 + 10 =$

(9) $9 + 10 =$

(10) $10 + 10 =$

3 Add.

4 points per question

(1) $1 + 9 =$

(2) $2 + 9 =$

(3) $4 + 10 =$

(4) $5 + 10 =$

(5) $8 + 10 =$

(6) $9 + 10 =$

(7) $10 + 10 =$

(8) $7 + 9 =$

(9) $8 + 9 =$

(10) $4 + 9 =$

Consonant Combinations

bl, cl, fl & pl

Level ★★

Score

Date　　/　　/

Name

/100

① Trace the words below. Color the boxes of the words with the same consonant blend to connect three in a row like tic-tac-toe.

25 points per question

(1) words that start with "bl"

block	float	clock
flag	blast	plum
plant	climb	black

(2) words that start with "cl"

block	clap	plane
fly	clean	flat
blimp	claw	plant

(3) words that start with "fl"

clock	plane	float
blimp	flute	blast
flap	plate	clue

(4) words that start with "pl"

play	plow	plant
clap	flap	blue
block	fly	clam

16

Addition

Level ★★

Score

/100

Math

DAY

9

Date   /   /

Name

1 Add.

(1) 7 + 1 =

(2) 9 + 1 =

(3) 6 + 1 =

(4) 6 + 2 =

(5) 8 + 2 =

(6) 5 + 2 =

(7) 6 + 3 =

(8) 5 + 3 =

(9) 1 + 4 =

(10) 3 + 4 =

(11) 4 + 5 =

(12) 7 + 5 =

2 Add.

(1) 1 + 6 =

(2) 2 + 6 =

(3) 3 + 6 =

(4) 4 + 7 =

(5) 5 + 7 =

(6) 3 + 7 =

(7) 4 + 8 =

(8) 5 + 8 =

(9) 6 + 8 =

(10) 6 + 9 =

(11) 5 + 9 =

(12) 2 + 9 =

(13) 3 + 9 =

(14) 3 + 10 =

(15) 5 + 10 =

(16) 7 + 10 =

Consonant Combinations
br, cr, fr & gr

Level ★★
Score

/10

Date / /   Name

① Trace the words below. Color the boxes of the words with the same consonant blend to connect three in a row like tic-tac-toe.

25 points per ques

(1) words that start with "br"

crib	frog	fry
grass	gray	crack
brick	brown	brain

(2) words that start with "cr"

green	fry	crab
brain	grin	cry
fruit	brick	crack

(3) words that start with "fr"

frog	grape	crawl
grab	fry	braid
cry	brag	fruit

(4) words that start with "gr"

grape	brown	crib
grass	fry	brick
green	crawl	fruit

Addition

Level ★★

Score

/100

Math

DAY

10

Date / /

Name

1 Add.

3 points per question

(1) $5 + 1 =$

(2) $5 + 2 =$

(3) $5 + 3 =$

(4) $4 + 5 =$

(5) $4 + 6 =$

(6) $4 + 7 =$

(7) $8 + 3 =$

(8) $8 + 4 =$

(9) $8 + 5 =$

(10) $7 + 7 =$

(11) $7 + 8 =$

(12) $7 + 9 =$

2 Add.

4 points per question

(1) $7 + 4 =$

(2) $8 + 3 =$

(3) $6 + 5 =$

(4) $9 + 6 =$

(5) $5 + 8 =$

(6) $2 + 7 =$

(7) $5 + 10 =$

(8) $3 + 9 =$

(9) $5 + 6 =$

(10) $2 + 4 =$

(11) $9 + 4 =$

(12) $7 + 1 =$

(13) $9 + 9 =$

(14) $8 + 5 =$

(15) $3 + 10 =$

(16) $4 + 5 =$

Reading
DAY
10

Consonant Combinations
sk, sm, sn & sp

Date / /

Name

Level ⭐⭐
Score

/1

1 Trace the words below. Color the boxes of the words with the same consonant blend to connect three in a row like tic-tac-toe.

25 points per quest

(1) words that start with "sk"

smart	smoke	space
sky	skip	skirt
spy	snow	snake

(2) words that start with "sm"

ski	smell	skip
snail	small	snow
spot	smile	speak

(3) words that start with "sn"

spoon	smoke	snake
small	ski	snow
space	skin	snap

(4) words that start with "sp"

snap	smell	spy
sky	spot	skirt
spoon	smart	snow

Two-Digit Addition

Level ★★★

Score

/100

Math

DAY

11

Date   /   /

Name

1 Add.

3 points per question

(1) $9 + 1 =$

(2) $10 + 1 =$

(3) $11 + 1 =$

(4) $10 + 2 =$

(5) $11 + 2 =$

(6) $12 + 2 =$

(7) $12 + 3 =$

(8) $13 + 3 =$

(9) $14 + 3 =$

(10) $12 + 4 =$

(11) $13 + 4 =$

(12) $14 + 4 =$

2 Add.

4 points per question

(1) $12 + 5 =$

(2) $13 + 5 =$

(3) $14 + 5 =$

(4) $14 + 6 =$

(5) $12 + 6 =$

(6) $13 + 6 =$

(7) $13 + 7 =$

(8) $12 + 7 =$

(9) $11 + 7 =$

(10) $11 + 8 =$

(11) $12 + 8 =$

(12) $10 + 8 =$

(13) $10 + 9 =$

(14) $11 + 9 =$

(15) $15 + 5 =$

(16) $11 + 6 =$

Reading
DAY
11

Consonant Combinations
gl, sl, st & tr

Date
/ /

Name

Level ★★
Score

/10

① Trace the words below. Color the boxes of the words with the 25 points per question same consonant blend to connect three in a row like tic-tac-toe.

（1）words that start with "gl"

truck	start	slip
glad	glove	glow
trip	slow	stew

（2）words that start with "sl"

train	slow	glad
step	slip	trail
stamp	sleep	glass

（3）words that start with "st"

slow	slug	start
trash	glad	step
tree	glue	stamp

（4）words that start with "tr"

step	glow	truck
stew	train	slow
tree	sleep	glove

Two-Digit Addition

1 Add.

3 points per question

(1) $12 + 2 =$

(2) $14 + 2 =$

(3) $16 + 2 =$

(4) $12 + 3 =$

(5) $14 + 3 =$

(6) $16 + 3 =$

(7) $12 + 4 =$

(8) $14 + 4 =$

(9) $16 + 4 =$

(10) $12 + 5 =$

(11) $14 + 5 =$

(12) $16 + 5 =$

2 Add.

4 points per question

(1) $12 + 6 =$

(2) $13 + 6 =$

(3) $14 + 6 =$

(4) $14 + 7 =$

(5) $12 + 7 =$

(6) $13 + 8 =$

(7) $14 + 8 =$

(8) $12 + 9 =$

(9) $17 + 4 =$

(10) $17 + 5 =$

(11) $17 + 7 =$

(12) $18 + 3 =$

(13) $18 + 5 =$

(14) $18 + 6 =$

(15) $19 + 3 =$

(16) $19 + 5 =$

If you want more addition practice, check out Kumon's *Addition Grade 1.*

Consonant Combinations
-nt, -mp, -nk & -st

Date / /

Name

① Complete each sentence by using words from the box below. 10 points per question

| elephant   jump   best   ink |

(1) I _____ high.

(2) An _____ has large ears.

(3) There is _____ on the desk.

(4) He was the _____ runner.

② Complete each sentence by using words from the box below. 15 points per question

| best ant tank stamp tent test lamp sink |

(1) The _____ went into the _____.

(2) The _____ shows a _____.

(3) We saw the _____ _____.

(4) I did my _____ on the _____.

Keep it up!

Vertical Addition

Date   /   /

Name

1 Add.

5 points per question

(1) $3 + 2 =$ ☐

(6)
$$\begin{array}{r} 3 \\ +\ 2 \\ \hline \square \end{array}$$
Write the answer here.

(11) $6 + 3 =$ ☐

(16)
$$\begin{array}{r} 6 \\ +\ 3 \\ \hline \square \end{array}$$

(2) $5 + 4 =$ ☐

(7)
$$\begin{array}{r} 5 \\ +\ 4 \\ \hline \square \end{array}$$

(12) $8 + 1 =$ ☐

(17)
$$\begin{array}{r} 8 \\ +\ 1 \\ \hline \square \end{array}$$

(3) $7 + 1 =$ ☐

(8)
$$\begin{array}{r} 7 \\ +\ 1 \\ \hline \square \end{array}$$

(13) $5 + 3 =$ ☐

(18)
$$\begin{array}{r} 5 \\ +\ 3 \\ \hline \square \end{array}$$

(4) $3 + 4 =$ ☐

(9)
$$\begin{array}{r} 3 \\ +\ 4 \\ \hline \square \end{array}$$

(14) $6 + 1 =$ ☐

(19)
$$\begin{array}{r} 6 \\ +\ 1 \\ \hline \square \end{array}$$

(5) $2 + 6 =$ ☐

(10)
$$\begin{array}{r} 2 \\ +\ 6 \\ \hline \square \end{array}$$

(15) $4 + 3 =$ ☐

(20)
$$\begin{array}{r} 4 \\ +\ 3 \\ \hline \square \end{array}$$

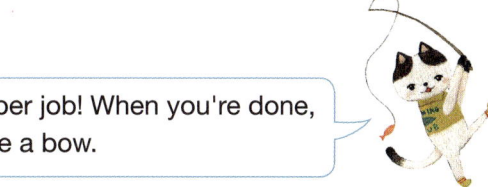

Super job! When you're done, take a bow.

Consonant Combinations
-sh, -ch, -nd & -ck

Date / /

Name

Level ★★
Score

/100

---

1 Complete each sentence by using words from the box below. 10 points per question

fish   lunch   truck   sand

(1) The _____ is here.

(2) I put my hand in the _____.

(3) The _____ swam away.

(4) She sat to eat her _____.

2 Complete each phrase by using words from the box below. 15 points per question

wash bench wand rock lunch duck dish hand

(1) I _____ the _____.

(2) We ate _____ on the _____.

(3) I held the _____ in my _____.

(4) The _____ sat on a _____.

Subtracting 1

Level ★

Score

/100

Math
DAY
14

Date   /   /

Name

1 Say the number sentence aloud as you subtract. Use the examples as hints.

5 points per question

(1) $2 - 1 = 1$

Two minus one equals one

(2) $3 - 1 = 2$

Three minus one equals two

(3) $4 - 1 =$

(4) $5 - 1 =$

(5) $6 - 1 =$

(6) $7 - 1 =$

(7) $8 - 1 =$

(8) $9 - 1 =$

(9) $10 - 1 =$

(10) $11 - 1 =$

2 Subtract.

5 points per question

(1) $2 - 1 =$

(2) $3 - 1 =$

(3) $5 - 1 =$

(4) $6 - 1 =$

(5) $4 - 1 =$

(6) $7 - 1 =$

(7) $9 - 1 =$

(8) $8 - 1 =$

(9) $10 - 1 =$

(10) $11 - 1 =$

1	2	3	4	5	6	7	8	9	10	11

Consonant Combinations
-ll, -ss, -ff, -mm, -nn & -rr

Level ★★
Score

Date   /   /

Name

/100

① Complete each sentence by using words from the box below.   10 points per question

| sniff   summer   carrot   class   funny   ill |

(1) I felt _____.

(2) He laughed at the _____ joke.

(3) The _____ is orange.

(4) We sat in the _____.

(5) The boy took a _____.

(6) In the _____ we sail boats.

② Complete each phrase by using words from the box below.   10 points per question

| offer bull sunny off summer grass tall berry |

(1) I will _____ some _____ juice.

(2) We walked in the _____ _____.

(3) It's _____ in the _____.

(4) The _____ threw the cowboy _____ his back.

If you want to more writing practice, check out Kumon's *My Book of WRITING WORDS*.

28

Level ⭐

Score

/100

Math
DAY
15

Date / /

Name

1 Say the number sentence aloud as you subtract. Use the examples as hints.

5 points per question

(1) 3 − 2 = 1

(2) 4 − 2 = 2

(3) 5 − 2 = 3

(4) 6 − 2 = 4

(5) 7 − 2 =

(6) 8 − 2 =

(7) 9 − 2 =

(8) 10 − 2 =

(9) 11 − 2 =

(10) 12 − 2 =

2 Subtract.

5 points per question

(1) 3 − 2 =

(2) 4 − 2 =

(3) 9 − 2 =

(4) 7 − 2 =

(5) 8 − 2 =

(6) 10 − 2 =

(7) 5 − 2 =

(8) 6 − 2 =

(9) 12 − 2 =

(10) 11 − 2 =

| 1 | 2 | 3 | 4 | 5 | 6 | 7 | 8 | 9 | 10 | 11 | 12 |

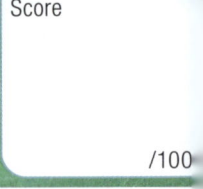

Reading DAY 15

Long Vowel Sounds
"a" & "e" sounds

Level ★★
Score

Date / /

Name

/100

① Pick the correct word from the box to match each picture below. 5 points per question

play   bee   mane   read   rain   meet

(1) _____

(2) _____

(3) _____

(4) _____

(5) _____

(6) _____

② Fill in the missing vowels. Use the pictures as clues. 5 points per question

(1) m _ n

(2) m _ _ t

(3) r _ _ d

(4) p l _ y

(5) r _ _ n

(6) b _ _

③ Trace the words below. Circle the word with the matching long vowel sound. 20 points per question

(1) Long "a" as in plane

mane   seat   fan

(2) Long "e" as in tree

 sky  bee  net

Date / /

Name

1 Say the number sentence aloud as you subtract. Use the examples as hints. 5 points per question

(1) $4 - 3 = 1$

(2) $5 - 3 = 2$

(3) $6 - 3 =$

(4) $7 - 3 =$

(5) $8 - 3 =$

(6) $9 - 3 =$

(7) $10 - 3 =$

(8) $11 - 3 =$

(9) $12 - 3 =$

(10) $13 - 3 =$

2 Subtract. 5 points per question

(1) $4 - 3 =$

(2) $5 - 3 =$

(3) $10 - 3 =$

(4) $11 - 3 =$

(5) $12 - 3 =$

(6) $8 - 3 =$

(7) $9 - 3 =$

(8) $6 - 3 =$

(9) $7 - 3 =$

(10) $13 - 3 =$

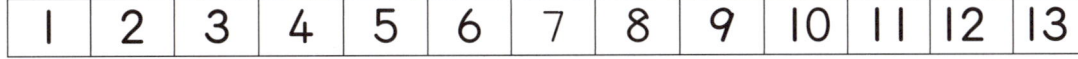

| 1 | 2 | 3 | 4 | 5 | 6 | 7 | 8 | 9 | 10 | 11 | 12 | 13 |

Long Vowel Sounds
"i" & "o" sounds

Date   /   /

Name

① Pick the correct word from the box to match each picture below.   5 points per question

| bite   close   bike   toad   slide   float |

(1) _____

(2) _____

(3) _____

(4) _____

(5) _____

(6) _____

② Fill in the missing vowels. Use the pictures as clues.   5 points per question

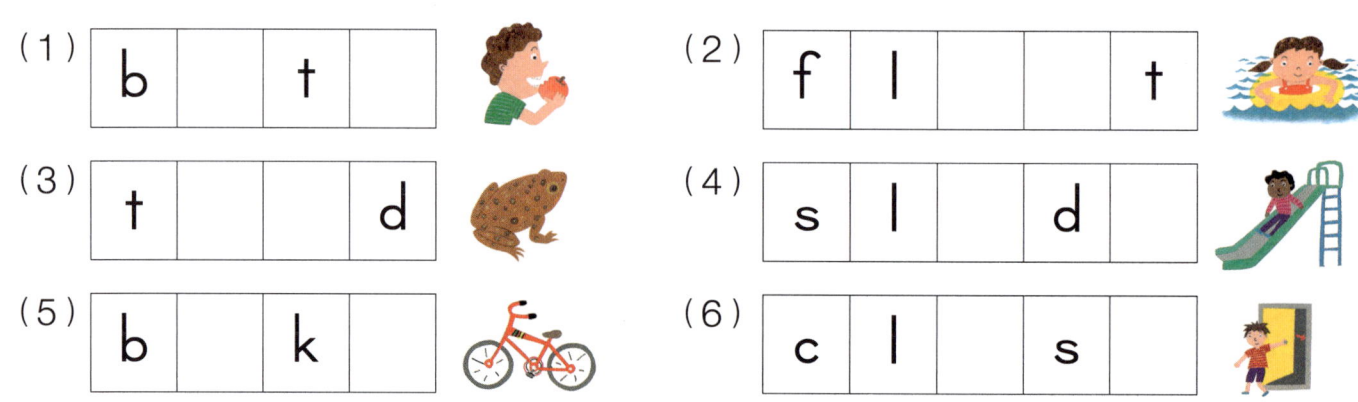

(1) | b | | | t |

(2) | f | l | | | t |

(3) | t | | | d |

(4) | s | l | | d |

(5) | b | | k |

(6) | c | l | | s |

③ Trace the words below. Circle the word with the matching long vowel sound.   20 points per question

(1) Long "i" as in kite

bite   goat   big

(2) Long "o" as in boat

log   cry   float

Subtracting 4

Level ★★
Score

/100

Math
DAY
17

1 Say the number sentence aloud as you subtract. Use the examples as hints. 5 points per question

(1) $5 - 4 = 1$

(2) $6 - 4 = 2$

(3) $7 - 4 =$

(4) $8 - 4 =$

(5) $9 - 4 =$

(6) $10 - 4 =$

(7) $11 - 4 =$

(8) $12 - 4 =$

(9) $13 - 4 =$

(10) $14 - 4 =$

2 Subtract. 5 points per question

(1) $5 - 4 =$

(2) $6 - 4 =$

(3) $10 - 4 =$

(4) $11 - 4 =$

(5) $12 - 4 =$

(6) $9 - 4 =$

(7) $8 - 4 =$

(8) $7 - 4 =$

(9) $14 - 4 =$

(10) $13 - 4 =$

1	2	3	4	5	6	7	8	9	10	11	12	13	14

33

Reading DAY 17

Long Vowel Sounds
"u" sounds & Review

Date / /

Name

Level ★★
Score

/1

① Pick the correct word from the box to match each picture 5 points per question
below.

| cute   glue   cube |

(1)  _____
(2) _____
(3) _____

② Fill in the missing vowels. Use the pictures as clues. 5 points per question

(1) | g | l | | |
(2) | c | | b | |
(3) | c | | t | |

③ Trace the words below. Circle the word with the matching long 10 points per question
vowel sound.

(1) Long "u" like flute

hide   cute   road

④ Complete each phrase by using a word from the box below. 15 points per question

| meet   plane   play   read |

(1) _____ in the rain
(2) _____ about a bee
(3) _____ on the street
(4) ride a _____ and a train

© Kumon Publishing Co.,Ltd.

34

Subtracting 5 & 6

Level ★★
Score

Math
DAY
18

/100

1 Subtract. Use the examples as hints.   3 points per question

(1) $6 - 5 = 1$

(2) $7 - 5 = 2$

(3) $8 - 5 =$

(4) $9 - 5 =$

(5) $10 - 5 =$

(6) $11 - 5 =$

(7) $12 - 5 =$

(8) $13 - 5 =$

(9) $14 - 5 =$

(10) $15 - 5 =$

2 Subtract. Use the example as a hint.   3 points per question

(1) $7 - 6 = 1$

(2) $8 - 6 =$

(3) $9 - 6 =$

(4) $10 - 6 =$

(5) $11 - 6 =$

(6) $12 - 6 =$

(7) $13 - 6 =$

(8) $14 - 6 =$

(9) $15 - 6 =$

(10) $16 - 6 =$

3 Subtract.   4 points per question

(1) $7 - 5 =$

(2) $8 - 5 =$

(3) $9 - 6 =$

(4) $10 - 6 =$

(5) $12 - 6 =$

(6) $12 - 5 =$

(7) $13 - 5 =$

(8) $15 - 6 =$

(9) $16 - 6 =$

(10) $14 - 5 =$

35

Long Vowel Sounds
Review

1 Complete each phrase by using a word from the box below.   10 points per question

| bike   toad   glue   slide   float   cube |

(1) _____ in a boat

(2) down the _____

(3) run, hike and _____

(4) _____ hops far

(5) put _____ on the blue part

(6) the tube around the _____

2 Read all the words aloud. Circle the word that matches each picture.   8 points per question

(1) 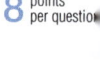   rain   mane   plane   play

(2)   tree   meet   read   street

(3)    kite   slide   wide   bite

(4)    toad   float   road   close

(5)    glue   tube   blue   cute

36

If you want more practice with long vowel sounds, check out Kumon's *My Book of RHYMING WORDS: LONG VOWELS.*

Subtracting 7 & 8

Level ★★

Score

/100

Math
DAY
19

Date / /

Name

1 Subtract. Use the examples as hints.

3 points per question

(1) $8 - 7 = 1$

(2) $9 - 7 = 2$

(3) $10 - 7 =$

(4) $11 - 7 =$

(5) $12 - 7 =$

(6) $13 - 7 =$

(7) $14 - 7 =$

(8) $15 - 7 =$

(9) $16 - 7 =$

(10) $17 - 7 =$

2 Subtract. Use the example as a hint.

3 points per question

(1) $9 - 8 = 1$

(2) $10 - 8 =$

(3) $11 - 8 =$

(4) $12 - 8 =$

(5) $13 - 8 =$

(6) $14 - 8 =$

(7) $15 - 8 =$

(8) $16 - 8 =$

(9) $17 - 8 =$

(10) $18 - 8 =$

3 Subtract.

4 points per question

(1) $10 - 7 =$

(2) $11 - 7 =$

(3) $10 - 8 =$

(4) $11 - 8 =$

(5) $8 - 7 =$

(6) $15 - 8 =$

(7) $16 - 8 =$

(8) $16 - 7 =$

(9) $14 - 7 =$

(10) $18 - 8 =$

Reading
DAY
19

Vocabulary
Months

Date / /

Name

Level ★★
Score

/10

① Trace each word below. Then read it aloud.   40 points for completion

(1) January

(2) February

(3) March

(4) April

(5) May

(6) June

(7) July

(8) August

(9) September

(10) October

(11) November

(12) December

② Complete the words by using the letters from the box below.   10 points per question
Hint: you can use the letters more than once.

| uary   ember   ober |

(1)  Jan_____

(2)  Febr_____

(3)  Sept_____

(4)  Oct_____

(5) Nov_____

(6) Dec_____

Date / /

Name

/100

1 Subtract. Use the example as a hint.

3 points per question

(1) $10 - 9 = 1$ (6) $15 - 9 =$

(2) $11 - 9 =$ (7) $16 - 9 =$

(3) $12 - 9 =$ (8) $17 - 9 =$

(4) $13 - 9 =$ (9) $18 - 9 =$

(5) $14 - 9 =$ (10) $19 - 9 =$

2 Subtract. Use the example as a hint.

3 points per question

(1) $11 - 10 = 1$ (6) $16 - 10 =$

(2) $12 - 10 =$ (7) $17 - 10 =$

(3) $13 - 10 =$ (8) $18 - 10 =$

(4) $14 - 10 =$ (9) $19 - 10 =$

(5) $15 - 10 =$ (10) $20 - 10 =$

3 Subtract.

4 points per question

(1) $10 - 9 =$ (6) $19 - 10 =$

(2) $11 - 9 =$ (7) $20 - 10 =$

(3) $14 - 10 =$ (8) $16 - 9 =$

(4) $15 - 10 =$ (9) $17 - 9 =$

(5) $18 - 10 =$ (10) $13 - 9 =$

39

Reading
DAY
20

Vocabulary
Homonyms

Date   /   /

Name

Level ★★
Score

/100

① Trace each word below. Connect the words that sound the same. **40** points for complet

(1)  whale •   • ⓐ wail

(2)  ball •   • ⓑ you

(3) ate •   • ⓒ scent

(4) ewe •   • ⓓ eight **8**

(5) cent •   • ⓔ bawl

(6) heel •   • ⓕ heal

② Circle the picture that matches each word below. **10** points per question

(1) bawl    (2) ewe

(3) scent     (4) ate

(5) whale     (6) heal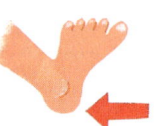

Subtraction

Date   /   /

Name

1 Subtract. Use the examples as hints.

4 points per question

(1) $6 - 1 = 5$

(2) $5 - 3 =$

(3) $9 - 3 =$

(4) $14 - 6 =$

(5) $8 - 6 =$

(6) $10 - 8 = 2$

(7) $7 - 3 =$

(8) $6 - 5 =$

(9) $13 - 7 =$

(10) $14 - 9 =$

(11) $15 - 6 =$

(12) $8 - 2 =$

(13) $13 - 8 =$

(14) $16 - 8 =$

(15) $11 - 4 =$

(16) $9 - 5 =$

(17) $15 - 7 =$

(18) $10 - 6 =$

(19) $14 - 8 =$

(20) $12 - 5 =$

(21) $3 - 1 =$

(22) $18 - 10 =$

(23) $12 - 6 =$

(24) $13 - 5 =$

(25) $17 - 9 =$

Vocabulary
Adjectives

Date / /

Name

① Trace the words below. Read each phrase aloud.   5 points per question

(1) _hot_ food

(2) _cold_ food

(3) _short_ rope

(4) _long_ rope

(5) _big_ cat

(6) _small_ kitten

(7) _new_ shoes

(8) _old_ shoes

(9) _slow_ bike

(10) _fast_ bike

② Finish each sentence with a word from the box that describes the picture.   10 points per question

slow   hot   new   short   big

(1) A snail is not fast. It is _____.

(2) The truck is not small. It is _____.

(3) The soup is not cold. It is _____.

(4) The elf is not tall. He is _____.

(5) The shoes are not old. They are _____.

Subtraction

Date   /   /

Name

1 Subtract.

4 points per question

(1) $14 - 8 =$

(2) $13 - 6 =$

(3) $11 - 9 =$

(4) $14 - 7 =$

(5) $16 - 9 =$

(6) $15 - 7 =$

(7) $17 - 8 =$

(8) $14 - 9 =$

(9) $12 - 5 =$

(10) $11 - 8 =$

(11) $10 - 2 =$

(12) $14 - 6 =$

(13) $16 - 8 =$

(14) $13 - 7 =$

(15) $12 - 4 =$

(16) $15 - 10 =$

(17) $18 - 9 =$

(18) $14 - 5 =$

(19) $10 - 7 =$

(20) $11 - 6 =$

(21) $13 - 8 =$

(22) $17 - 9 =$

(23) $15 - 6 =$

(24) $12 - 3 =$

(25) $16 - 7 =$

Your work is out of sight!

Vocabulary
Verbs

Date   /   /

Name

Level ★★
Score

/100

① Trace the words below. Read each sentence aloud.

5 points per question

(1) I _eat_ .

(2) I _sleep_ .

(3) I _walk_ .

(4) I _run_ .

(5) I _climb_ .

(6) I _fall_ .

(7) I _spin_ .

(8) I _jump_ .

(9) I _play_ .

(10) I _swim_ .

② Complete each sentence with the correct verb from the box below.

10 points per question

walk	swim	jump	climb	sleep

(1) Lions _____.

(2) Monkeys _____.

(3) Kangaroos _____.

(4) Bears _____.

(5) Penguins _____.

Subtraction

Level ★★★

Score

/100

Math
DAY
23

Date
/   /

Name

1 Subtract.

3 points per question

(1) $10 - 1 =$

(2) $11 - 1 =$

(3) $12 - 1 =$

(4) $11 - 2 =$

(5) $12 - 2 =$

(6) $13 - 2 =$

(7) $12 - 3 =$

(8) $13 - 3 =$

(9) $14 - 3 =$

(10) $13 - 4 =$

(11) $14 - 4 =$

(12) $15 - 4 =$

2 Subtract.

4 points per question

(1) $14 - 5 =$

(2) $15 - 5 =$

(3) $16 - 5 =$

(4) $16 - 6 =$

(5) $17 - 6 =$

(6) $19 - 6 =$

(7) $16 - 7 =$

(8) $17 - 7 =$

(9) $18 - 7 =$

(10) $20 - 7 =$

(11) $17 - 8 =$

(12) $18 - 8 =$

(13) $20 - 8 =$

(14) $18 - 9 =$

(15) $19 - 9 =$

(16) $20 - 9 =$

If you want more subtraction practice, check out Kumon's *Subtraction Grade 1*.

Vocabulary
Compound Words

Date   /   /

Name

① Read each word. Then trace the compound word and read it aloud.   10 points per question

(1) foot + ball = football

(2) blue + berry = blueberry

(3) class + room = classroom

(4) gold + fish = goldfish

(5) sand + castle = sandcastle

② Make words to match the pictures below by connecting two puzzle pieces.   10 points per question

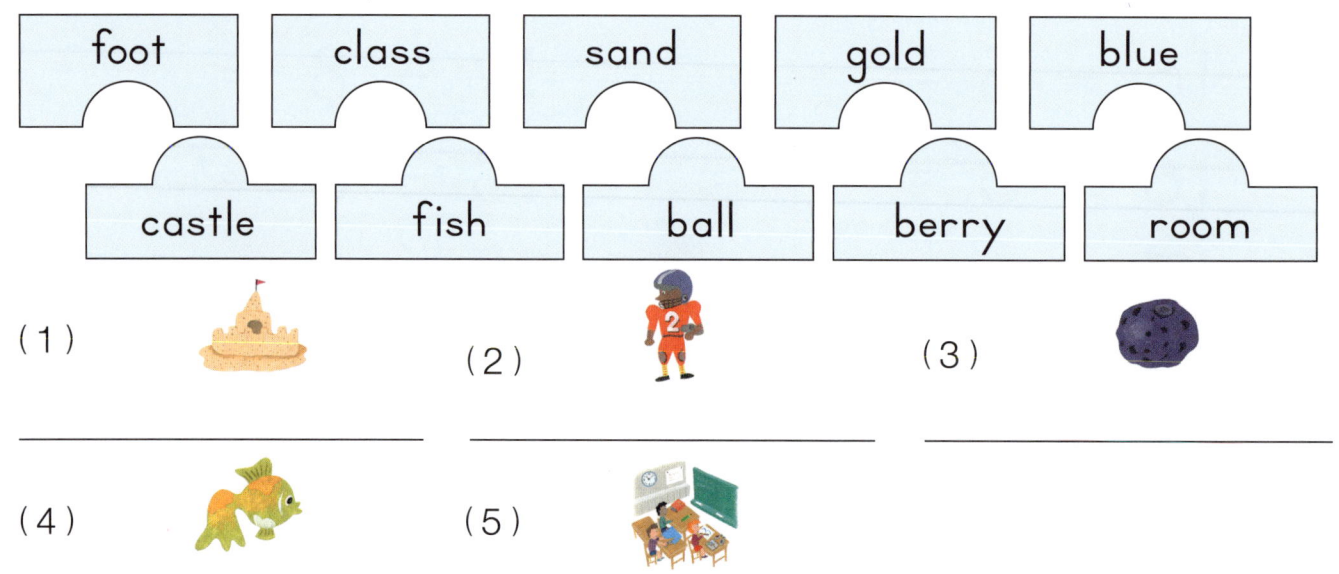

| foot | class | sand | gold | blue |

| castle | fish | ball | berry | room |

(1) _____

(2) _____

(3) _____

(4) _____

(5) _____

Vertical Subtraction

Date / /   **Name**

1 Subtract.

5 points per question

(1) $9 - 3 = \boxed{}$

(6)
$$\begin{array}{r} 9 \\ - \ 3 \\ \hline \boxed{} \end{array}$$
☞ Write the answer here.

(11) $10 - 6 = \boxed{}$

(16)
$$\begin{array}{r} 10 \\ - \ 6 \\ \hline \boxed{} \end{array}$$

(2) $10 - 3 = \boxed{}$

(7)
$$\begin{array}{r} 10 \\ - \ 3 \\ \hline \boxed{} \end{array}$$

(12) $10 - 8 = \boxed{}$

(17)
$$\begin{array}{r} 10 \\ - \ 8 \\ \hline \boxed{} \end{array}$$

(3) $10 - 5 = \boxed{}$

(8)
$$\begin{array}{r} 10 \\ - \ 5 \\ \hline \boxed{} \end{array}$$

(13) $11 - 3 = \boxed{}$

(18)
$$\begin{array}{r} 11 \\ - \ 3 \\ \hline \boxed{} \end{array}$$

(4) $9 - 4 = \boxed{}$

(9)
$$\begin{array}{r} 9 \\ - \ 4 \\ \hline \boxed{} \end{array}$$

(14) $10 - 7 = \boxed{}$

(19)
$$\begin{array}{r} 10 \\ - \ 7 \\ \hline \boxed{} \end{array}$$

(5) $10 - 4 = \boxed{}$

(10)
$$\begin{array}{r} 10 \\ - \ 4 \\ \hline \boxed{} \end{array}$$

(15) $10 - 9 = \boxed{}$

(20)
$$\begin{array}{r} 10 \\ - \ 9 \\ \hline \boxed{} \end{array}$$

If you want more subtraction practice, check out Kumon's *Subtraction Grade 2*.

Vocabulary

Contractions

Date / /

Name

① Read each word. Then trace the contraction. 40 points for complete

(1) can + not = _can't_

(2) could + not = _couldn't_

(3) is + not = _isn't_

(4) was + not = _wasn't_

(5) do + not = _don't_

(6) did + not = _didn't_

② Read the contraction. Then write the two words that each 10 points per question
contraction represents.

(1) isn't = _____ + _____

(2) didn't = _____ + _____

(3) couldn't = _____ + _____

(4) wasn't = _____ + _____

(5) can't = _____ + _____

(6) don't = _____ + _____

Word Problems

Addition

Date	Name
/ /	

Level ★★

Score

/100

Math

DAY

25

1 Read the word problem and write the number sentence below. Then answer the question. 20 points per question

(1) You have 3 sheets of red paper and 2 sheets of blue paper. How many sheets of paper do you have?

$$\boxed{3} + \boxed{2} = \boxed{}$$

Ans. $\boxed{}$ sheets

(2) There are 4 books on your desk and 2 books on your bookshelf. How many books are there?

$$\boxed{4} + \boxed{} = \boxed{}$$

Ans. $\boxed{}$ books

(3) There are 5 pencils in your pencil case and 3 pencils in your backpack. How many pencils are there?

$$\boxed{} + \boxed{} = \boxed{}$$

Ans. $\boxed{}$ pencils

(4) 6 cars are parked. 3 more cars park. How many cars are parked?

$$\boxed{} + \boxed{} = \boxed{}$$

Ans. $\boxed{}$ cars

(5) You have 5 fish in your pond. You put in 2 more fish. How many fish are in your pond?

$$\boxed{} + \boxed{} = \boxed{}$$

Ans. $\boxed{}$ fish

Date / /

Name

① Trace the words. Then draw a line between the two words that are synonyms, or have the same meaning.

10 points per question

(1) fat •

• ⓐ ship

(2) warm •

• ⓑ dad

(3) boat •

• ⓒ kind

(4) father •

• ⓓ plump

(5) nice •

• ⓔ toasty

② Read the sentences. From the box below, choose the synonym of the underlined word and complete the sentence.

10 points per question

kind   plump   ship   dad   toasty

(1) The blanket is <u>warm</u>. I am _____.

(2) The cat is <u>fat</u>. It has a _____ belly.

(3) My <u>father</u> is fun. I love my _____.

(4) My teacher is <u>nice</u>. I also try to be _____.

(5) Is that <u>boat</u> strong? A _____ must be sturdy.

Word Problems
Addition

Date	Name
/ /	

Level ★★
Score

/100

Math
DAY
26

1 Read the word problem and write the number sentence below. Then answer the question. 20 points per question

(1) There are 3 cookies on the dish and 4 cookies in the box. How many cookies are there in all?

3 + 4 =

Ans. _____ cookies

(2) 6 children are playing on the slide and 3 children are playing on the swings. How many children are there in all?

Ans. _____

(3) 4 birds are eating food. 2 more birds join. How many birds are there in all?

Ans. _____

(4) You have 5 stickers. If I give you 3 stickers, how many stickers will you have in all?

Ans. _____

(5) There are 7 apples on the table. If you put 2 more apples on the table, how many apples will there be in all?

Ans. _____

Antonyms

Date / /

Name

/1

① Trace the words. Then draw a line between the two words that are antonyms, or have opposite meanings. 10 points per ques

(1) short  • • ⓐ low

(2) light  • • ⓑ long

(3) dirty  • • ⓒ go

(4) stop  • • ⓓ heavy

(5) high  • • ⓔ clean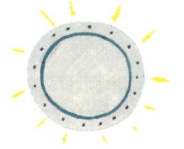

② Read the sentences. From the box below, choose the antonym of the underlined word and complete the sentence. 10 points per quest

| heavy   low   clean   short   go |

(1) When a cat is <u>dirty</u>, he licks his fur _____.

(2) A brick is _____, but a feather is <u>light</u>.

(3) A rollercoaster climbs <u>high</u> and drops ____.

(4) Don't <u>stop</u>! Just ____!

(5) A squirrel tail is <u>long</u>. A bunny tail is _____.

Word Problems
Addition

Date / /

Name

Level ★
Score

/100

Math
DAY
27

1 Read the word problem and write the number sentence below. 20 points per question
Then answer the question.

(1) You had 7 books. Then you bought 4 books. How many books do you have in all?

Ans.

(2) You want to send 8 cards to your friends and 2 cards to your family. How many cards do you need?

Ans.

(3) There are 3 sheets of blue paper. There are 4 more sheets of green paper than blue paper. How many sheets of green paper are there?

Ans.

(4) There are 6 flowerpots. There are 3 more bulbs than flowerpots. How many bulbs are there?

Ans.

(5) Diana ate 5 strawberries. Her sister ate 6 more strawberries. How many strawberries did her sister eat?

Ans.

Hooray for you! Job well done!

Reading
DAY
27

Irregular Plurals

Level ⭐⭐
Score

/10

Date   /   /   Name

① Complete the chart below by tracing the plural words. Note how each word changes when it is plural, or is more than one in number.   20 points for comp

One	More than one	One	More than one
beach	beaches	puppy	puppies
brush	brushes	man	men
wish	wishes	goose	geese
tornado	tornadoes	child	children

② Complete each sentence using a word from the box.   5 points per question

> geese puppies children men

(1) The _____ look nice in their suits.

(2) Look at the _____ flying south.

(3) I love to pet _____.

(4) All _____ like games.

③ Read the sentences. Add "es" to the underlined words to complete the sentence.   15 points per question

(1) A painter has brush___.

(2) People make wish___ at a fountain.

(3) Seals swim to different beach___.

(4) Scientists watch tornado___.

Word Problems
Subtraction

Date / /

Name

Level ⭐
Score
/100

Math
DAY
28

1 Read the word problem and write the number sentence below. **20** points per question
Then answer the question.

(1) There were 6 candies on the table. You ate 3 candies. How many candies were left?

6 − 3 = ☐

Ans. ☐ candies

(2) There are 10 pencils on your desk. You give 4 pencils away. How many pencils are left?

10 − ☐ = ☐

Ans. ☐ pencils

(3) There were 8 sheets of paper. You used 5 sheets of paper for coloring. How many sheets of paper are left?

☐ − ☐ = ☐

Ans. ☐ sheets

(4) There are 7 eggs. You eat 3 eggs. How many eggs are left?

☐ − ☐ = ☐

Ans. ☐ eggs

(5) There were 9 sparrows on the roof. 2 sparrows flew away. How many were left?

☐ − ☐ = ☐

Ans. ☐ sparrows

Who, When & Where

Date
/ /

Name

① Read the passage. Then complete the exercises below.

In the spring, Rabbit and Pig plant a garden. First, Rabbit picks a spot in the yard. Then Pig rakes the soil and drops the seeds. Then Rabbit hops around to push the seeds down. Finally, Pig gets the hose from the house. He waters the garden.

(1) Circle the correct answers to each question.

10 points per question

ⓐ Who picks a spot?

ⓑ Who rakes the soil?

ⓒ Who hops around?

ⓓ Who waters the garden?

(2) Choose a word from the box to answer the questions.

15 points per question

| last   house   spring   yard |

ⓐ When do Rabbit and Pig plant a garden?

   They plant a garden in the _____ .

ⓑ Where do Rabbit and Pig plant a garden?

   They plant a garden in the _____.

ⓒ When does Pig water the garden?

   Pig waters the garden _____.

ⓓ Where does Pig get the hose?

   Pig gets the hose from the _____.

Word Problems

Subtraction

Date
/   /

Name

Level ★★
Score

/100

Math
DAY
29

1 Read the word problem and write the number sentence below. 20 points per question
Then answer the question.

(1) Kate had 9 dollars. She used 3 dollars to buy a snack. How many dollars does she have left?

Ans. _____ dollars

(2) Emmy has 8 oranges. If she eats 2 oranges, how many oranges will she have left?

Ans. _____

(3) There are 7 baseballs and 4 footballs. How many more baseballs than footballs are there?

Ans. _____

(4) There are 10 jump ropes and 5 hula hoops. How many more jump ropes than hula hoops are there?

Ans. _____

(5) There are 9 cats and 4 dogs. How many more cats than dogs are there?

Ans. _____

Reading
DAY
29

Who, When & Where

Date   /   /

Name

---

① Read the passage. Then complete the exercises below.

Pig waters the garden in the yard all spring. The fruits and vegetables are ready to eat in the summer. The corn is ripe in June. The tomatoes are ready in July. The berries are sweet in August. Pig and Rabbit pick the fruits and vegetables. They eat many meals with each other in the kitchen.

ⓐ Use a word from the passage to answer the questions.   15 points per question

(1) When is the corn ripe?

The corn is ripe in _____.

(2) When are the berries sweet?

The berries are sweet in _____ .

(3) When are the tomatoes ready?

The tomatoes are ready in _____.

(4) When does Pig water the garden?

Pig waters the garden all _____

ⓑ Circle the correct answer to each question.   10 points per question

(1) When are the plants ready to eat?

summer   fall   winter

(2) Where do Pig and Rabbit eat?

kitchen   yard   bath

(3) When does Pig water the garden?

summer   fall   spring

(4) Where is the garden?

kitchen   yard   street

Word Problems

Subtraction

Level ★★

Score

Math
DAY

30

/100

Date
/   /

Name

1 Read the word problem and write the number sentence below.
Then answer the question.

(1) Our class has 8 boys. We have 2 fewer girls than boys. How many girls are there?

Ans. _____ girls

(2) Mike has 10 baseball cards. He has 3 fewer toy cars than baseball cards. How many toy cars does he have?

Ans. _____

(3) There are 6 chairs. If 4 children each sit on a chair, how many chairs are left?

Ans. _____

(4) There are 8 cookies and 5 dishes. If you want to put 1 cookie on each dish, how many more dishes will you need?

Ans. _____

(5) The teacher had 10 sheets of paper. She gave 1 sheet of paper each to 7 children. How many sheets of paper does she have now?

Ans. _____

Who, When & Where

Date / /

Name

① **Read the passage. Use words from the passage to answer the** 20 points per question **questions below.**

Leah loves to dance. She goes to dance class on Monday. The teacher shows her a new dance. Leah works on it each day after school in her room. On Sunday, she shows her dance to her mom and dad in the living room. Her parents love it and clap.

(1) Who loves to dance?

_____ **loves to dance.**

(2) Where does Leah work on the dance?

Leah works on it in her _____.

(3) When does Leah go to dance class?

Leah goes to dance class on _____.

(4) Who does Leah show her dance to?

She shows her dance to her _____ and _____.

(5) When does Leah show her dance?

She shows her dance on _____.

Word Problems
Addition or Subtraction

Level ★★
Score

/100

Math
DAY
31

Date

/   /

Name

1 Read the word problem and write the number sentence below. 20 points per question
Then answer the question.

(1) There are 8 cows and 10 sheep at the farm. How many more sheep than cows are there?

Ans. _____

(2) Jim is 7 years old. His brother is 4 years older than him. How old is his brother?

Ans. _____

(3) There are 9 bottles of milk. 3 of them are empty. How many full bottles of milk are there?

Ans. _____

(4) There are 6 frogs in the pond. There are 5 more fish than frogs. How many fish are there in the pond?

Ans. _____

(5) Grandmother has 13 grandchildren. 7 of them are girls. How many boys are there?

Ans. _____

Great job! You are a champ!

Sequence

Date /   /

Name

① Read the passage and look at the pictures. Then number the 100 points for comp
pictures in the order in which they happened.

> April Fool's Day is the first day of April. I like to play a joke on my brother. First, I get a box. I put a stone in it to make the box heavy. I glue the top on and let it dry. Then, I wrap the box in nice paper and a bow. I give it to my brother. When he can't open it, I say, "April Fool's!"

ⓐ

_____

ⓑ

_____

ⓒ

_____

ⓓ

_____

ⓔ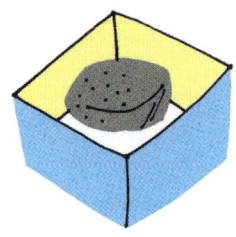

_____

ⓕ

_____

Word Problems

Addition or Subtraction

Level ★★

Score

Math

DAY

32

Date
/   /

Name

/100

1 Read the word problem and write the number sentence below. Then answer the question.

20 points per question

(1) The teacher has 8 pencils. There are 15 children in the class. How many children won't get pencils?

Ans. _____

(2) In the parking lot, there are 7 bikes. There are 6 more cars than bikes. How many cars are in the parking lot?

Ans. _____

(3) There are 10 carrots and 6 peppers on the table. Are there more carrots or peppers? How many more?

Ans.   There are _____ more _____ .

(4) There are 13 boys and 9 girls in the class. Are there fewer boys or girls? How many fewer?

Ans.   There are _____ fewer _____ .

(5) Mary has 17 stamps. If she uses 8 stamps, how many stamps are left?

Ans. _____

True or False

Date   /   /

Name

/10

① Read the passage. Then read the sentences below. Circle the "T" if the sentence is true, or correct. Circle the "F" if the sentence is false, or wrong.

10 points per question

Mara and Aunt Bee bake cupcakes. They need butter, flour, sugar and eggs. First, Aunt Bee breaks the eggs. Mara pours the sugar. They add the butter and flour and stir it all in a bowl. Mara uses a spoon to drop the batter in the cupcake pan. Aunt Bee puts the pan in the oven. When the cupcakes are done, Mara puts icing on top.

(1) Aunt Bee is baking with Mara.   T   F

(2) Mara puts the icing on top first.   T   F

(3) Aunt Bee pours the sugar.   T   F

(4) Aunt Bee breaks the eggs.   T   F

(5) They need apples for the cupcakes.   T   F

(6) They stir it all in a bowl.   T   F

(7) Mara uses a fork to drop the batter.   T   F

(8) Mara and Aunt Bee bake potatoes.   T   F

(9) Aunt Bee puts the pan in the oven.   T   F

(10) Mara pours the flour first.   T   F

Telling Time

Level ★★

Score

/100

Math
DAY
33

Date / /

Name

---

1 What time is it? Write the time under each clock.

5 points per question

(1)

(6:00)

(6)

(   )

(11)

(   )

(16)
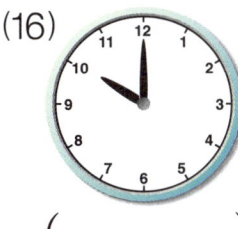
(   )

(2)
(1:30)

(7)

(   )

(12)
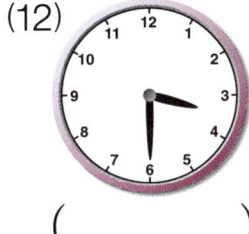
(   )

(17)

(   )

(3)
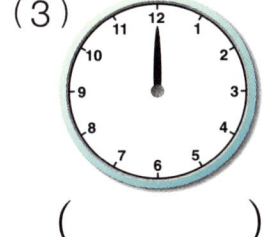
(   )

(8)
(   )

(13)

(   )

(18)

(   )

(4)
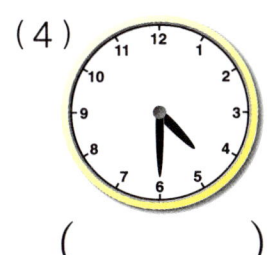
(   )

(9)
(   )

(14)

(   )

(19)
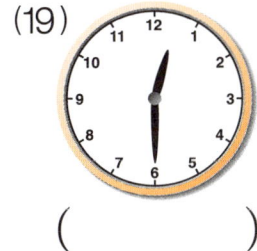
(   )

(5)
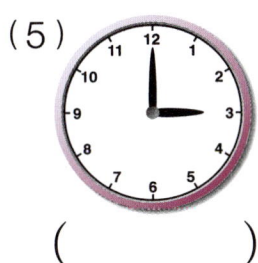
(   )

(10)
(   )

(15)

(   )

(20)
(   )

Don't forget! The short hand points to the hour. The long hand points to the minutes. When the long hand is pointed at 12, it means it's the start of the hour.

True or False

Date / /

Name

① Read the passage. Then read the sentences below. Circle the "T" if the sentence is true, or correct. Circle the "F" if the sentence is false, or wrong.

10 points per quest

> Fox goes to the fair with his mom. They eat cotton candy. Fox rides the Ferris wheel. He also wants to play a game. Fox and his mom pick the milk bottle game. They each get three balls to throw. They must knock down three bottles to win. Fox knocks down two bottles. Fox's mom knocks all three bottles down. She gets a prize and gives it to Fox. It is a toy airplane!

(1) Fox goes to the zoo with his mom.　　　　T　　　　F

(2) Fox and his mom eat cotton candy.　　　　T　　　　F

(3) Fox rides the Ferris wheel.　　　　T　　　　F

(4) Fox wants to play in the fun house.　　　　T　　　　F

(5) Fox and his mom pick the milk bottle game.　　　　T　　　　F

(6) They each get ten balls to throw.　　　　T　　　　F

(7) They eat ice cream.　　　　T　　　　F

(8) Fox knocks all the bottles down.　　　　T　　　　F

(9) The prize is a toy airplane.　　　　T　　　　F

(10) They must knock down five bottles to win.　　　　T　　　　F

Wow! You've got it now!

Date   /   /

Name

1 What time is it? Write the time under each clock.

(1)
( )

(6)
( )

(11)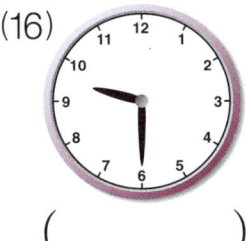
( )

(16)
( )

(2)
( )

(7)
( )

(12)
( )

(17)
( )

(3)
( )

(8)
( )

(13)
( )

(18)
( )

(4)
( )

(9)
( )

(14)
( )

(19)
( )

(5)
( )

(10)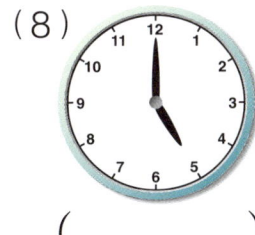
( )

(15)
( )

(20)
( )

If you want more practice telling time, check out Kumon's *My Book of Easy Telling Time* and *My Book of Telling Time*.

Who, What, Where & When

Level ★★★
Score

/1

Date   /   /

Name

① Read the passage. Use a word from the box to complete the questions below. Hint: You can use the words more than once.

20 points per quest

Some poems can be made into songs. Sarah Josepha Hale wrote the poem "Mary Had a Little Lamb" a long time ago. The poem was about a real girl named Mary and a lamb. Mary took the lamb to school as a joke. Lowell Mason made the poem into a song. He wrote the tune a few years after Sarah wrote the poem.

Who	What	Where	When

(1) __What__ **can be made into songs?**
Some poems can be made into songs.

(2) _____ **wrote the poem "Mary Had a Little Lamb?"**
Sarah Josepha Hale wrote the poem "Mary Had a Little Lamb."

(3) _____ **did Lowell Mason do?**
Lowell Mason made the poem into a song.

(4) _____ **did Mary take the lamb?**
Mary took the lamb to school.

(5) _____ **did Lowell Mason write the tune?**
Lowell Mason wrote the tune a few years after Sarah wrote it.

Counting Coins

Level ★★
Score

Math
DAY

35

/100

Date   /   /

Name

1 Write the value of each coin in the box.

5 points per question

(1)  **1** ¢   I penny   (5) **10** ¢   I dime

(2)  ☐ ¢   I penny   (6) ☐ ¢   I dime

(3)  **5** ¢   I nickel   (7)  **25** ¢   I quarter

(4)  ☐ ¢   I nickel   (8)  ☐ ¢   I quarter

2 Add the value of each group of coins. Then write the amount in the box on the right.

10 points per question

(1)  ☐ ¢

(4)  ☐ ¢

(2)  ☐ ¢

(5)  ☐ ¢

(3)  ☐ ¢

(6)  ☐ ¢

Reading
DAY
35
Who, What, Where, When & Why

Level ★★★
Score

/10

Date   /   /

Name

① Read the passage. Use a word from the box to complete the questions below.   20 points per question

> We study tornadoes at school. A tornado is air that spins fast. Tornadoes strike most often in spring. They start when warm, wet air mixes with cool, dry air. They can wreck homes and cars. If there is a tornado, you must go to a cellar. A cellar is safe since it is under the ground. You may need a flashlight because the power can go out.

Who	What	Where	When	Why

(1) _____ is a tornado?
A tornado is air that spins.

(2) _____ do tornadoes strike most often?
Tornadoes strike most often in spring.

(3) _____ must you go if there is a tornado?
You must go to a cellar if there is a tornado.

(4) _____ studies tornadoes?
We study tornadoes.

(5) __Why__ do tornadoes start?
Tornadoes start because warm, wet air mixes with cool, dry air.

Level ⭐⭐

Score

/100

Date / /

Name

1 Add the value of each group of coins. Then write the amount in the box on the right.

10 points per question

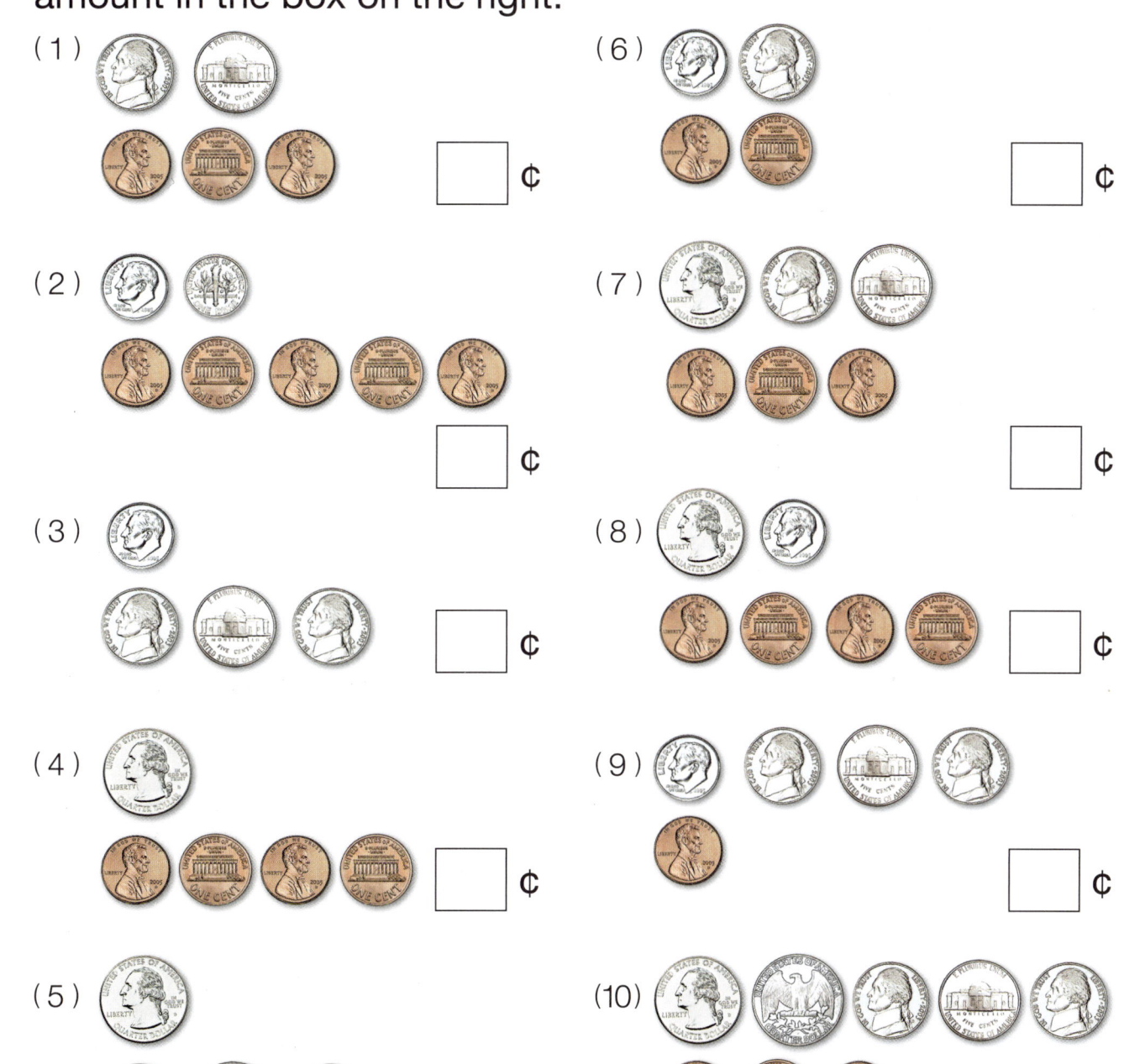

(1) ____ ¢

(2) ____ ¢

(3) ____ ¢

(4) ____ ¢

(5) ____ ¢

(6) ____ ¢

(7) ____ ¢

(8) ____ ¢

(9) ____ ¢

(10) ____ ¢

If you want more practice counting coins, check out Kumon's *My Book of Money: Counting Coins*.

71

Reading a Book Cover

Date
/ /

Name

① Read the following book cover. Then answer the questions below using words from the covers.　100 points for co

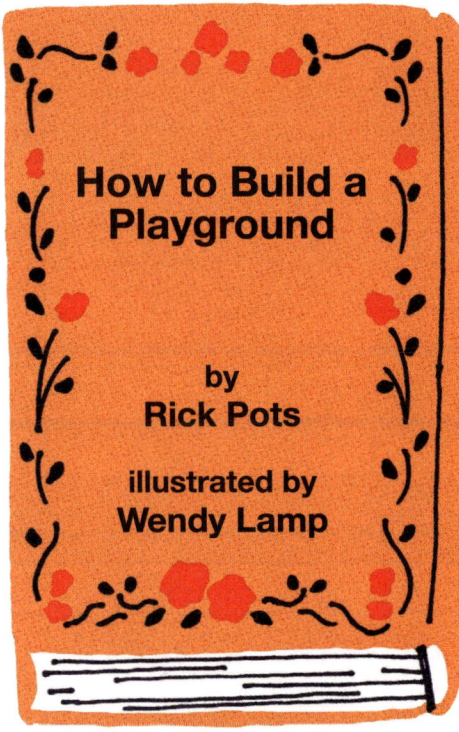

How to Build a
Playground

by
Rick Pots

illustrated by
Wendy Lamp

(1) What is the title, or name, of the book?
The title of the book is,
How to Build a _____.

(2) Who is the author, or person who wrote the book?
The author is Rick _____ .

(3) Who is the illustrator, or person who drew the pictures in the book?
The illustrator is Wendy _____.

(4) What is the title, or name, of the book?
The title of the book is
The _____ *of the Koala Twins.*

(5) Who is the author, or person who wrote the book?
The author is Sia _____.

(6) Who is the illustrator, or person who drew pictures in the book?
The illustrator is Ameya _____.

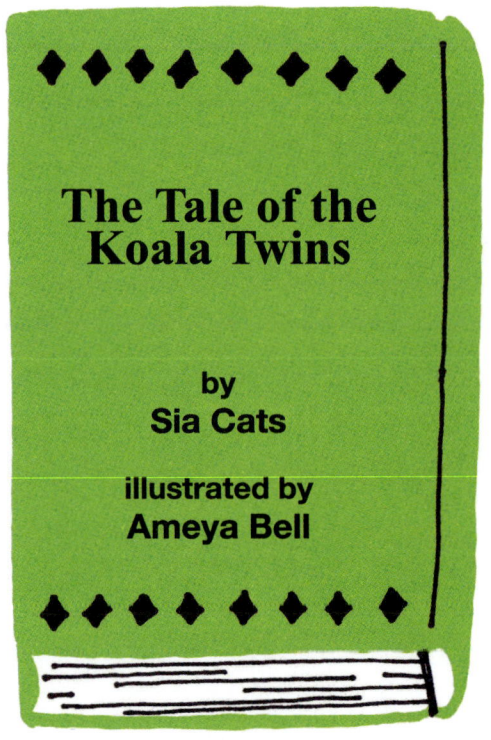

The Tale of the
Koala Twins

by
Sia Cats

illustrated by
Ameya Bell

Length

Level ★★

Score

Math

DAY

37

/100

Date   /   /

Name

1 Compare the length of the two objects shown and put a check (✓) next to the longer one.

15 points per question

(1) ⓐ
      ( )

      ⓑ
      ( )

(2) ⓐ
      ( )

      ⓑ
      ( )

(3) ⓐ
      ( )

      ⓑ
      ( )

(4) ⓐ
      ( )

      ⓑ
      ( )

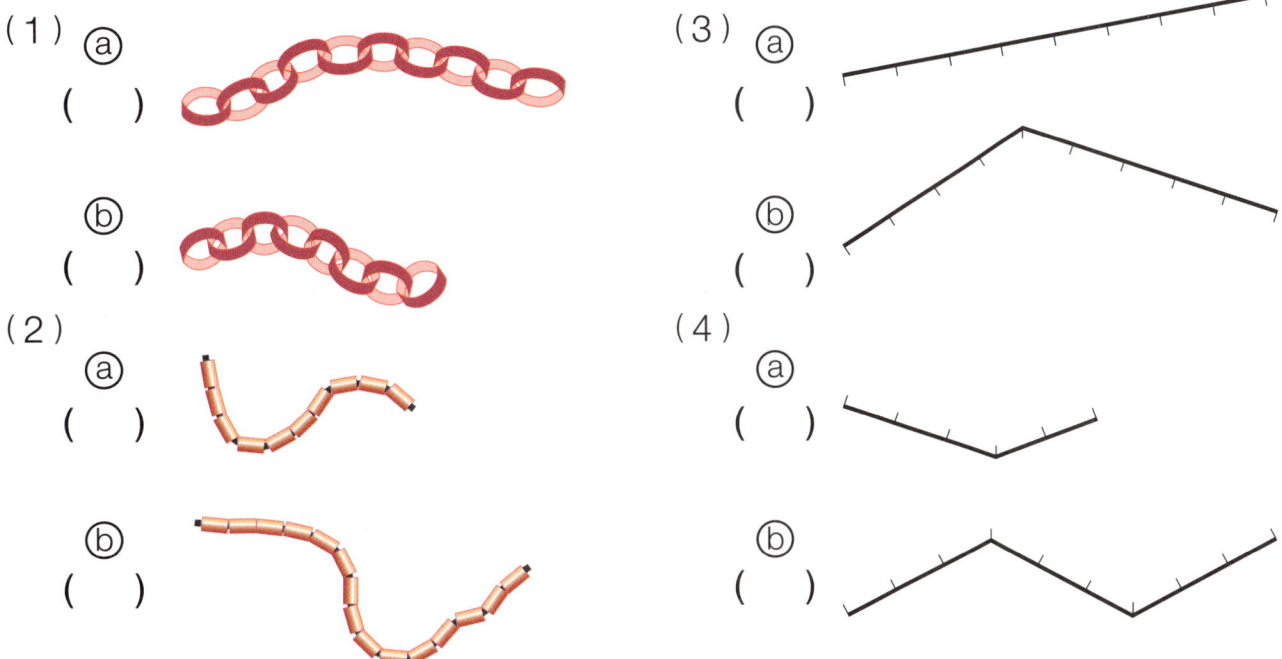

2 Compare the length of the items below and then answer the questions.

20 points per question

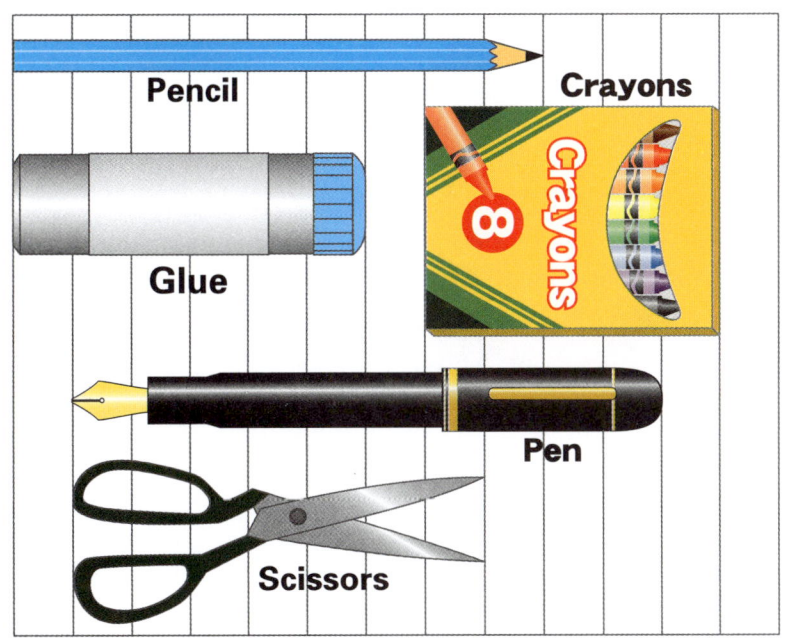

Pencil

Crayons

Glue

Pen

Scissors

(1) Which item is the longest?

(   )

(2) Which item is the shortest?

(   )

Reading a Table of Contents

Date   /   /

Name

① Read the following Table of Contents. Then answer the 20 points per question questions below.

Pirate Pete and the Dolphin Team

Chapter	Page
Sail Away	2
Pirate Pete Sleeps Under the Stars	8
A Storm is Coming	12
Dolphin Team Blows the Alarm	17
The Big Wave	22
Pirate Pete Falls in the Sea	26
Dolphin Team Dives	30
Pirate Pete Saved	34
Land Ahead	38

(1) What is the title, or name, of this book?

The title of this book is _____ Pete and the Dolphin _____.

(2) What page does the story start on?

The story starts on page _____.

(3) What is the name of the first chapter?

The name of the first chapter is " _____ Away."

(4) What page does "The Big Wave" start on?

"The Big Wave" starts on page _____ .

(5) Based on this Table of Contents, do you think Pirate Pete is saved by the Dolphin Team?

_____.

This is tough. Keep up the good work!

Length

Date / /

Name

Level ★★

Score /100

Math
DAY
38

1 Use the graph paper behind the book to answer questions about the length and width of this book. Each box on the paper is 1 unit.

20 points per question

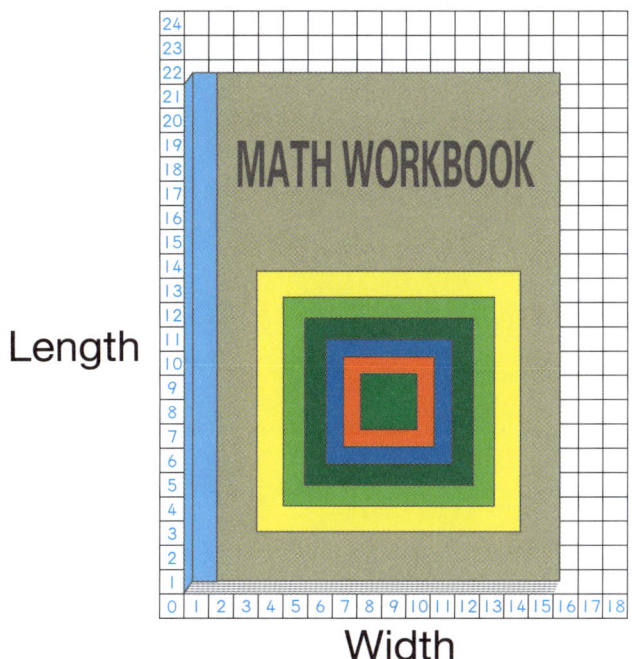

Length

Width

(1) The length is **21** units.

(2) The width is ☐ units.

(3) The length is ☐ units longer than the width.

2 How many inches is each arrow from the left side of the ruler to each box?

20 points per question

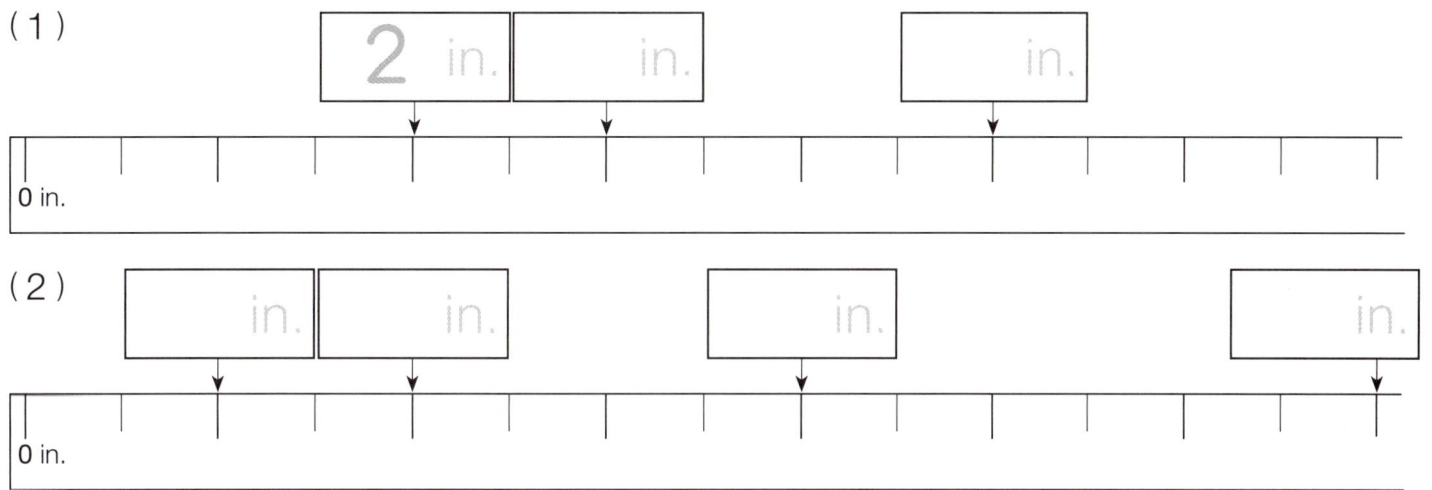

(1)

2 in.   in.   in.

0 in.

(2)

in.   in.   in.   in.

0 in.

Reading Comprehension

Date   /   /

Name

① Read the article below. Use words from the passage to answer the questions.

20 points per question

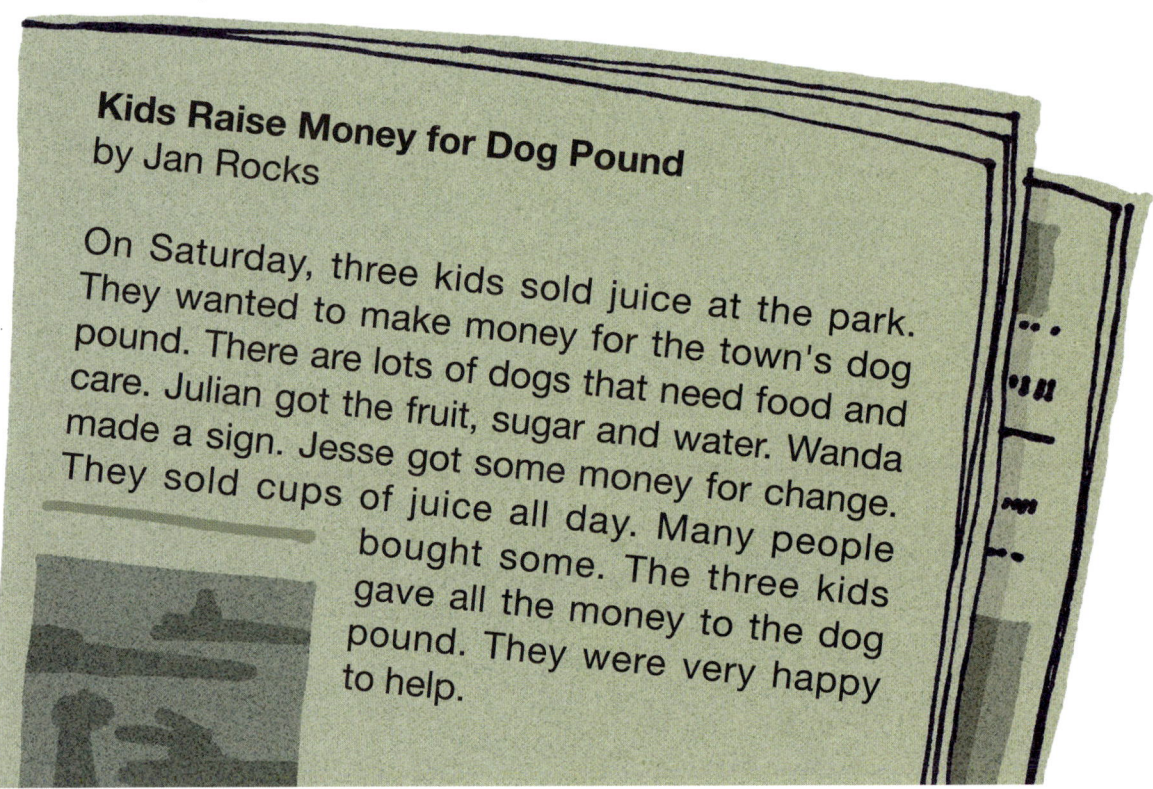

Kids Raise Money for Dog Pound
by Jan Rocks

On Saturday, three kids sold juice at the park. They wanted to make money for the town's dog pound. There are lots of dogs that need food and care. Julian got the fruit, sugar and water. Wanda made a sign. Jesse got some money for change. They sold cups of juice all day. Many people bought some. The three kids gave all the money to the dog pound. They were very happy to help.

(1) Who sold juice?

Three _____ sold juice?.

(2) What did Julian get?

Julian got the _____ , _____ and _____.

(3) Where did they sell cups of juice?

They sold cups of juice at the _____ .

(4) When did the three kids sell juice?

The three kids sold juice on _____ .

(5) Why did they sell juice?

They sold juice to make _____ for the town's dog pound.

Weight

Level ★★

Score

/100

Math
DAY
39

Date   /   /

Name

1 Which object is heavier? Circle the heavier object.

10 points per question

(1)

(2)

2 All of the blocks below are the same size and weight. Circle the heavier group of blocks.

10 points per question

(1)

(2)

3 How many cubes does the object on the right side of the scale weigh?

12 points per question

(1)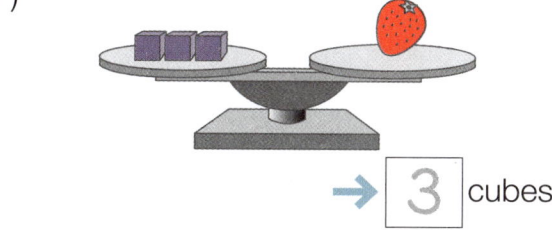

→ ⎡3⎤ cubes

(4)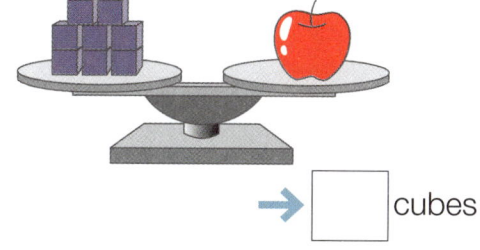

→ ⎡ ⎤ cubes

(2)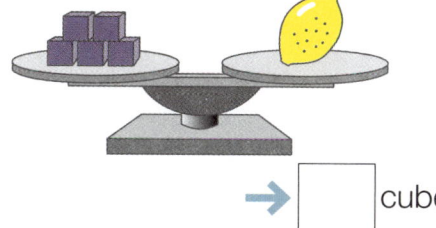

→ ⎡ ⎤ cubes

(5)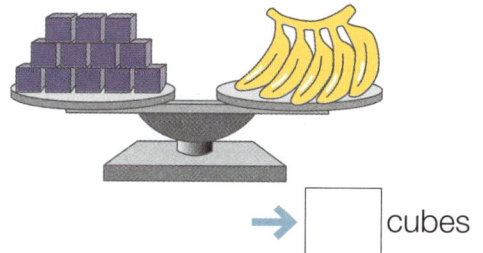

→ ⎡ ⎤ cubes

(3)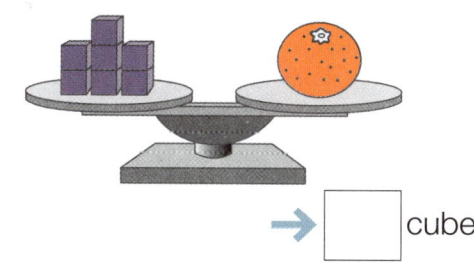

→ ⎡ ⎤ cubes

Amazing work! Keep it up!

Reading
DAY
39

Reading Comprehension

Level ★★★
Score

/100

Date / /

Name

1 Read the article below. Use words from the passage to answer the questions.

20 points per question

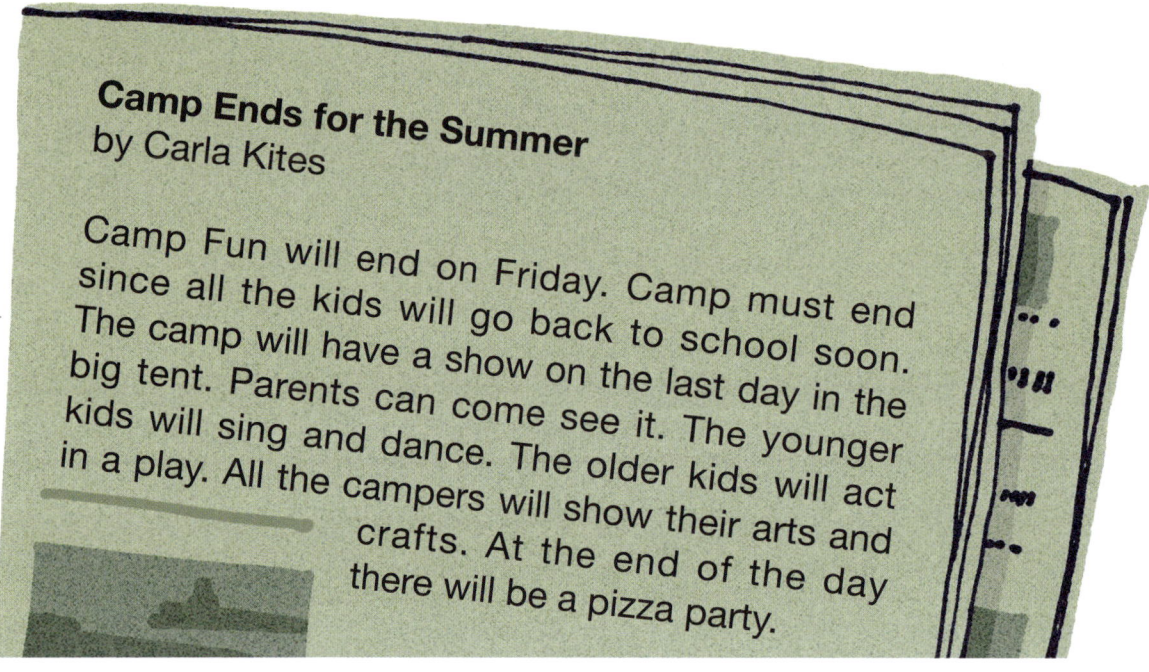

Camp Ends for the Summer
by Carla Kites

Camp Fun will end on Friday. Camp must end since all the kids will go back to school soon. The camp will have a show on the last day in the big tent. Parents can come see it. The younger kids will sing and dance. The older kids will act in a play. All the campers will show their arts and crafts. At the end of the day there will be a pizza party.

(1) When will Camp Fun end?
Camp Fun will end on _____ .

(2) Where will the camp have a show?
The camp will have a show in the big _____.

(3) Who can come to see the camp's show?
The _____ can come to see the camp's show.

(4) What will the younger kids do?
The younger kids will _____ and _____.

(5) Why must camp end?
Camp must end since all the kids will go back to _____ soon.

Weight

Level ★★★

Score

Date
/   /

Name

/100

Math
DAY
40

1 Read the weight on each scale and write it below.

10 points per question

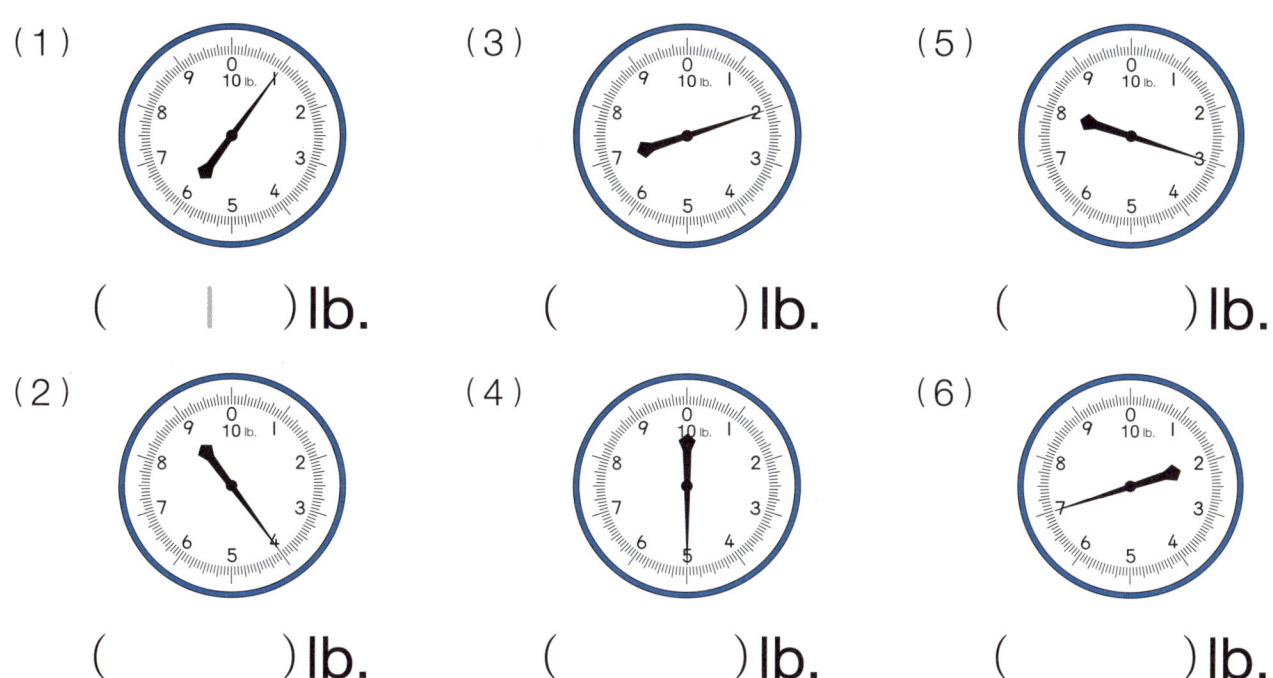

(1)   (3)   (5)

(|)lb.   (   )lb.   (   )lb.

(2)   (4)   (6)

(   )lb.   (   )lb.   (   )lb.

2 Read the weight on each scale and write it below.

8 points per question

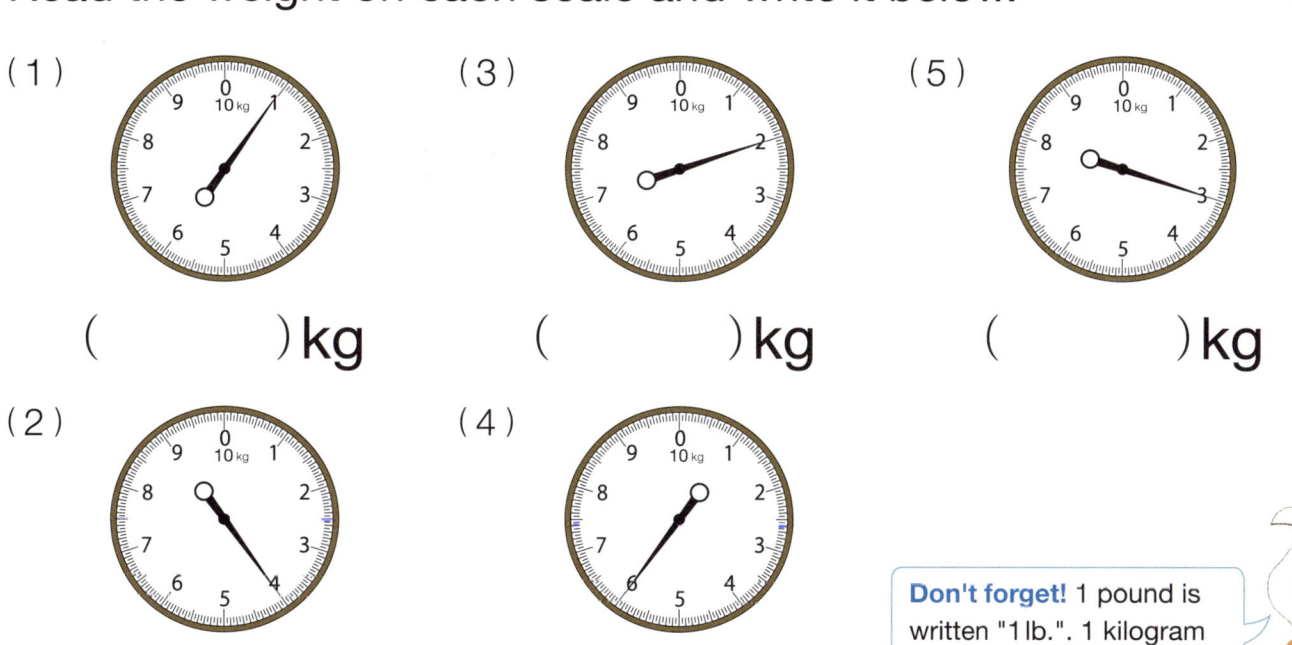

(1)   (3)   (5)

(   )kg   (   )kg   (   )kg

(2)   (4)

(   )kg   (   )kg

Don't forget! 1 pound is written "1 lb.". 1 kilogram is written "1 kg".

Reading Comprehension
The Best Art 1

Level ★★★
Score

Date
/ /

Name

/10

① Read the passage. Use words from the passage to answer   20 points per question
the questions below.

It was the last day of school. Ike was packing his bag.
He packed his pencils and books away. His teacher came
to pass out the students' final drawings. The teacher
gave Ike his drawing. Ike had drawn a big zoo with lions
and apes. He wanted to show it to his mom and dad. He
planned to put it up in his room. Ike put the drawing in his
bag with care.
   Ike's friend, Paula, also got her drawing from the teacher.
She didn't look happy. She put it in the trash.

（1）Who packed his bag?

_____ packed his bag.

（2）What did Ike pack?

Ike packed _____ and _____.

（3）Where did Paula put her drawing?

She put it in the _____.

（4）When did Ike pack his bag?

Ike packed his bag on the last day of _____.

（5）Why did Ike want to bring his drawing home?

Ike wanted to show his drawing to his _____ and _____ , and put it
up in his _____ .

Shapes

Date   /   /

Name

Level ★★

Score

/100

Math DAY 41

1 Connect the objects below to the shapes that are similar.   20 points per question

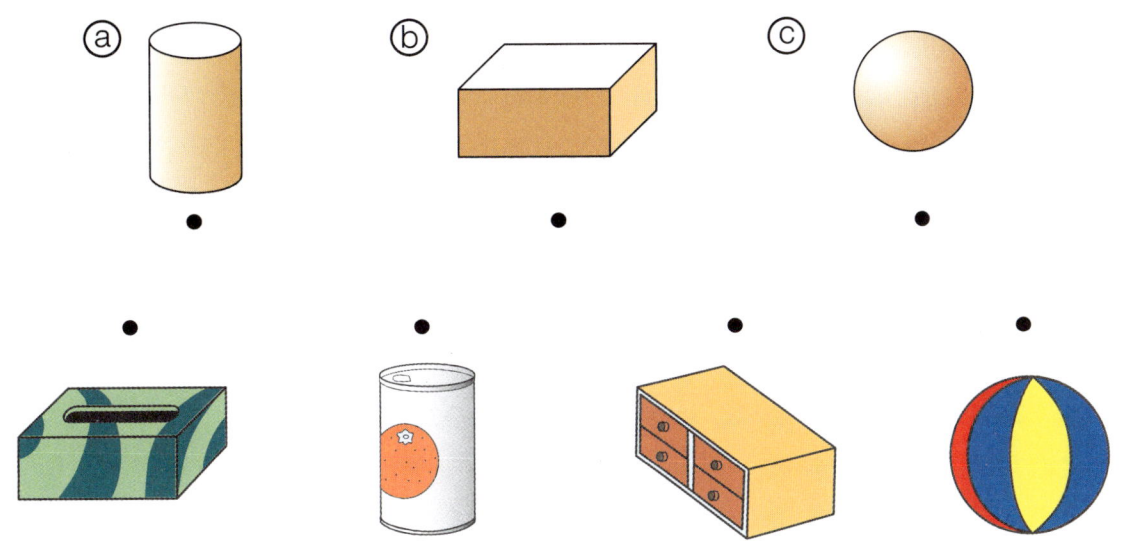

ⓐ   ⓑ   ⓒ

2 The picture below shows some blocks. How many of each block are used?   10 points per question

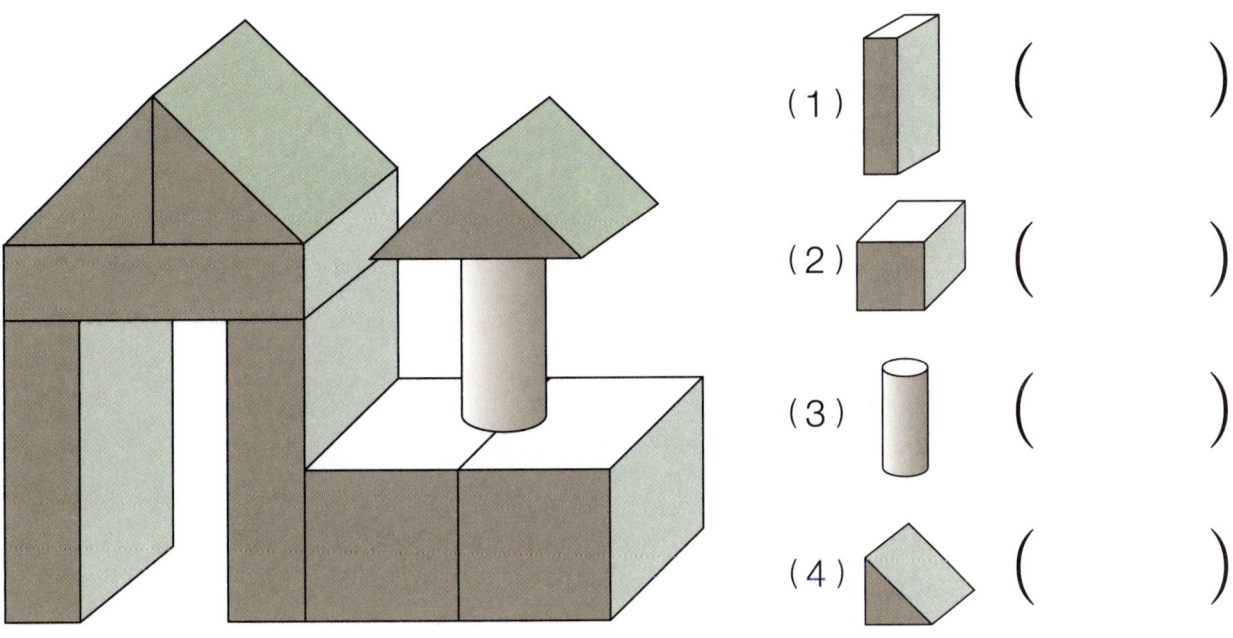

(1) (   )

(2) (   )

(3) (   )

(4) (   )

© Kumon Publishing Co.,Ltd.

81

Reading Comprehension
The Best Art 2

Date   /   /

Name

① Read the passage. Use words from the passage to answer   **20** points per question
the questions below.

After Paula put her drawing in the trash, she began to pack her bag. Ike asked, "Why did you put your drawing in the trash?"

"No one said that they liked it," Paula replied.

Paula left the classroom. Ike took the drawing from the trash because he wanted to see it. It was a map of the town. Paula had used crayons to color all the houses. Ike found his own house at the top of the map. He liked the map. He put it in his bag.

(1) Who had used crayons to color all the houses?

_____ **had used crayons to color all the houses.**

(2) What was the drawing?

The drawing was a map of the _____.

(3) Where did Ike find his house on the map?

Ike found his house at the _____ of the map.

(4) When did Ike pull the drawing from the trash?

Ike pulled the drawing from the trash after Paula left the _____.

(5) Why did Ike take the drawing from the trash?

Ike took the drawing from the trash because he wanted to_____ it.

Shapes

Level ★★
Score

Math
DAY
42

/100

1 How many triangles ▲ are used to create the figures below? 10 points per question

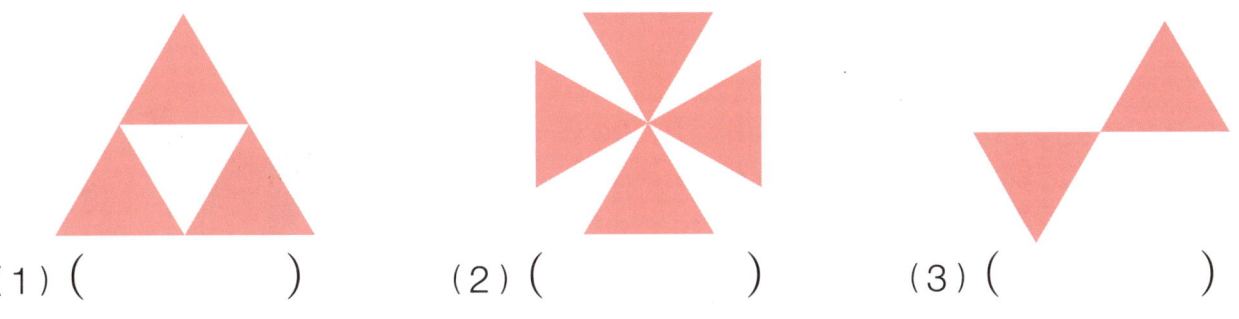

(1) (   )   (2) (   )   (3) (   )

2 You used three triangles (▲) to create the figures below. 10 points per question
Draw two lines to separate the three triangles in each shape.

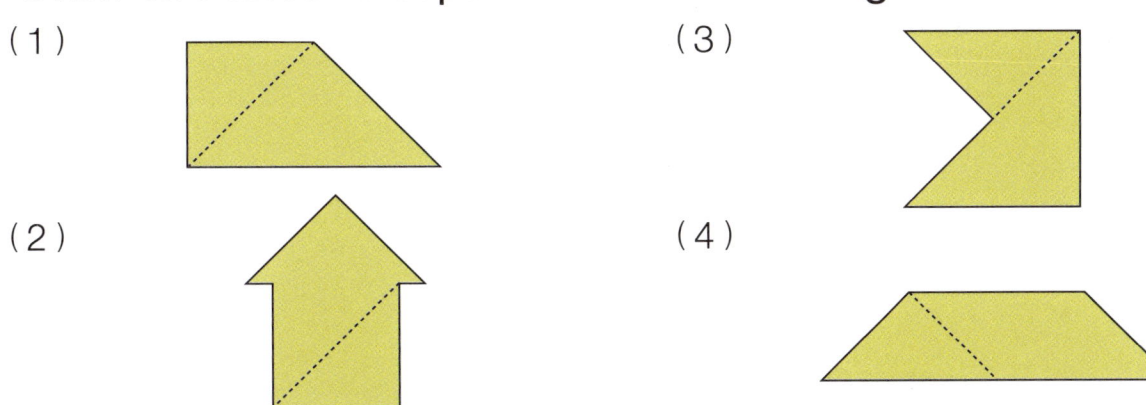

(1)

(2)

(3)

(4)

3 Draw lines to separate the same-size triangles in each shape. 15 points per question
How many same-size triangles were used to create the figures
below?

(1)

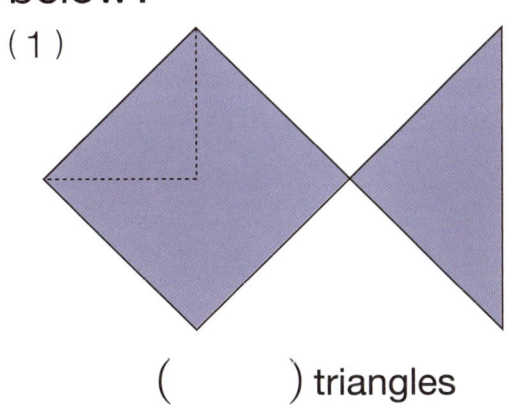

(   ) triangles

(2)

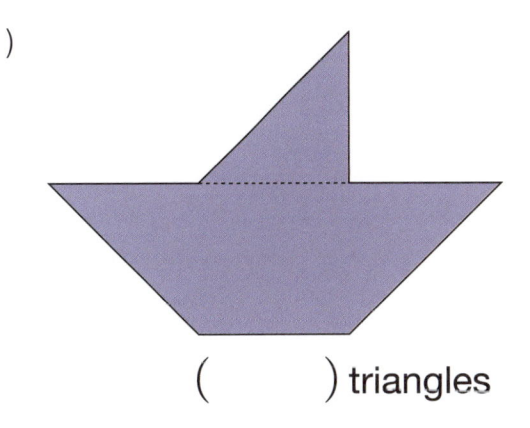

(   ) triangles

83

If you want more geometry practice,
check out Kumon's *Geometry &
Measurement Grade 1 and Grade 2*.

Reading Comprehension
The Best Art 3

Level ★★★
Score

Date   /   /

Name

/1

① Read the passage. Use words from the passage to answer the questions below.

20 points per que

Paula went to Ike's house to play the next day. Ike asked her, "Did you like your drawing?"

"Yes, I did." Paula said.

"If you like it, that is all that matters!" Ike said.

"Do you think so? Well, I wish I had it again. I am very sorry I put it in the trash," Paula replied sadly.

Ike pulled the map out of his bag and gave it to Paula. Paula was so happy that Ike saved her drawing.

"Thank you! You are right. If I like it, that is all that matters."

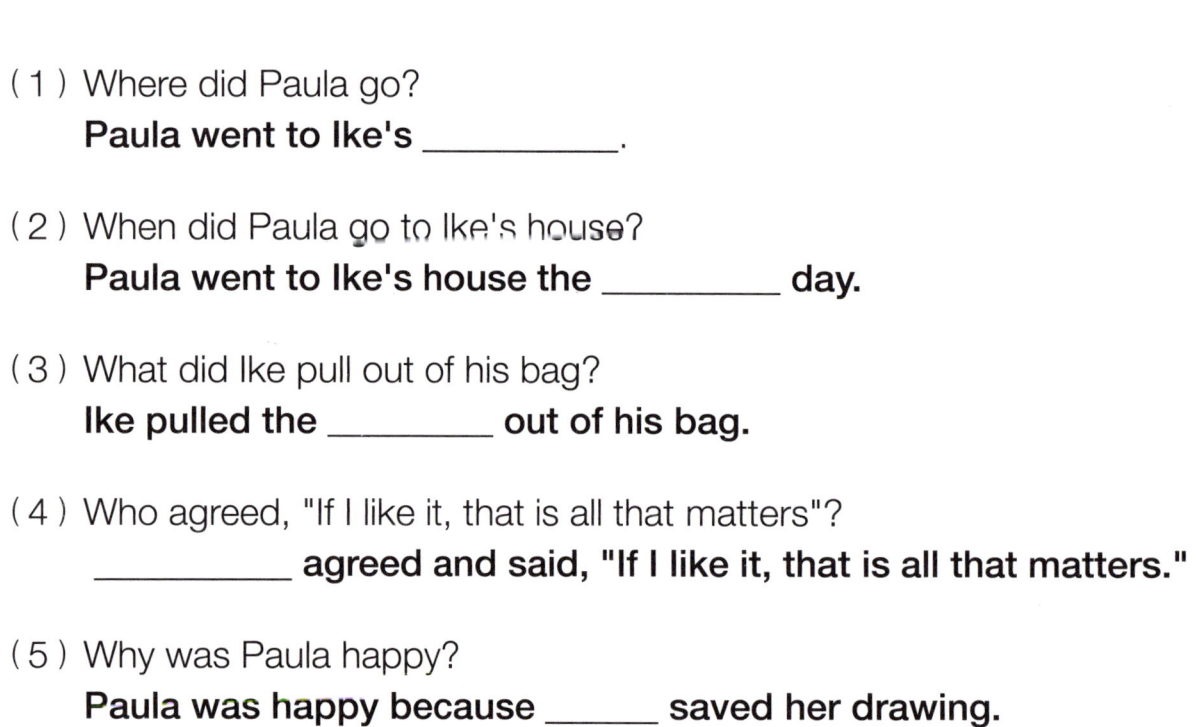

(1) Where did Paula go?

Paula went to Ike's _____.

(2) When did Paula go to Ike's house?

Paula went to Ike's house the _____ day.

(3) What did Ike pull out of his bag?

Ike pulled the _____ out of his bag.

(4) Who agreed, "If I like it, that is all that matters"?

_____ agreed and said, "If I like it, that is all that matters."

(5) Why was Paula happy?

Paula was happy because _____ saved her drawing.

If you want more reading practice, check out Kumon's *Reading Grade 1 and Grade 2.*

84

Review

Level ★★

Score

/100

Math

DAY

43

Date / /

Name

1 Add or subtract.

5 points per question

(1) $9 + 2 =$

(2) $11 - 5 =$

(3) $15 + 3 =$

(4) $13 - 2 =$

(5) $8 + 6 =$

(6) $14 - 6 =$

(7) $7 + 4 =$

(8) $10 - 4 =$

(9) $9 + 8 =$

(10) $13 - 8 =$

(11) $14 + 5 =$

(12) $15 - 5 =$

2 Read the word problem and write the number sentence below. Then answer the question.

10 points per question

(1) There are 5 candies on the dish and 7 candies in the box. How many candies are there in all?

Ans. _____

(2) There are 9 chairs. If 6 children sit on chairs, how many are left?

Ans. _____

(3) You have 8 stickers. If I give you 3 stickers, how many stickers will you have in all?

Ans. _____

(4) There are 10 apples and 6 oranges. How many more apples than oranges are there?

Ans. _____

Review

Level ★★★
Score

Date / /

Name

/1

① Write the correct vowels to finish each word below. Use the pictures as hints.   6 points per question

(1) m_ _n

(2) j_t

(3) l_p

(4) j_b

(5) h_g

(6) f_n

(7) p_g

(8) b_g

(9) l_g

(10) c_p

② Complete each sentence using a word from the box below.   5 points per question

> buffalo wish tent wand lock stamp lunch bunch

(1) You can _____ upon a star.

(2) Don't _____ yourself out of the house.

(3) Put a _____ on the envelope.

(4) Our _____ was yummy.

(5) We put the _____ up.

(6) I grabbed a _____ of crayons.

(7) They watched the _____ in the field.

(8) The fairy held a _____ .

You're almost at the end and you've made great progress!

Review

Level ★★

Score

/100

Math

DAY

44

Date   /   /

Name

1 Add or subtract.

5 points per question

(1) $12 - 5 =$

(2) $11 - 7 =$

(3) $13 + 2 =$

(4) $10 - 4 =$

(5) $11 - 3 =$

(6) $10 - 2 =$

(7) $10 + 5 =$

(8) $15 + 4 =$

(9) $14 + 1 =$

(10) $12 - 8 =$

(11) $11 - 6 =$

(12) $13 - 3 =$

2 Read the word problem and write the number sentence below. Then answer the question.

10 points per question

(1) Julie is 7 years old. Her brother is 6 years older than her. How old is her brother?

Ans. _____

(2) There are 14 boys and 9 girls in a class. Are there fewer boys or girls? How many fewer?

Ans. There are _____ fewer _____ .

3 What time is it? Write the time under each clock.

5 points per question

(1)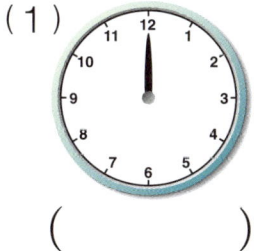

(   )

(2)

(   )

(3)

(   )

(4)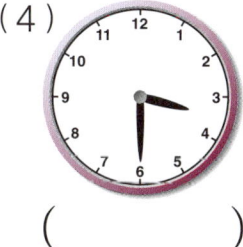

(   )

Reading
DAY
44

Review

Date / /

Name

Level ★★★
Score

/100

1 Fill in the missing vowels. Use the pictures as clues.

5 points per question

(1) p l __ y

(2) c __ b __

(3) k __ t __

(4) r __ __ d

(5) r __ __ n

(6) m __ __ t

(7) t __ __ d

(8) g l __ __

(9) b __ __ t

(10) b __ t

2 Finish each sentence with a word from the box that describes the picture.

10 points per question

| long | heavy | clean | old | fast |

(1) My hair is not short.

   It is _____.

(2) The car is not new.

   It is _____.

(3) The dish is not dirty.

   It is _____.

(4) This book is not light.

   It is _____.

(5) A lion is not slow.

   It is _____.

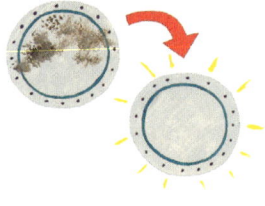

Review

Level ★★

Score

/100

Math

DAY

45

Date   /   /

Name

1 Add the value of each group of coins. Then write the amount in the box on the right.

10 points per question

(1)

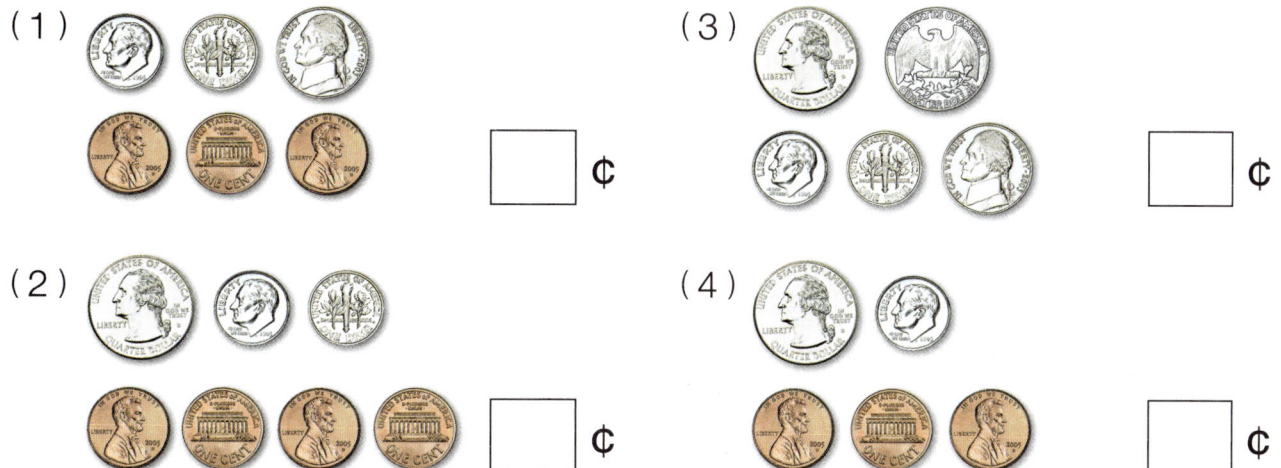

☐ ¢

(3)

☐ ¢

(2)

☐ ¢

(4)

☐ ¢

2 How many inches is it from the left side of the ruler to each arrow?

20 points for completion

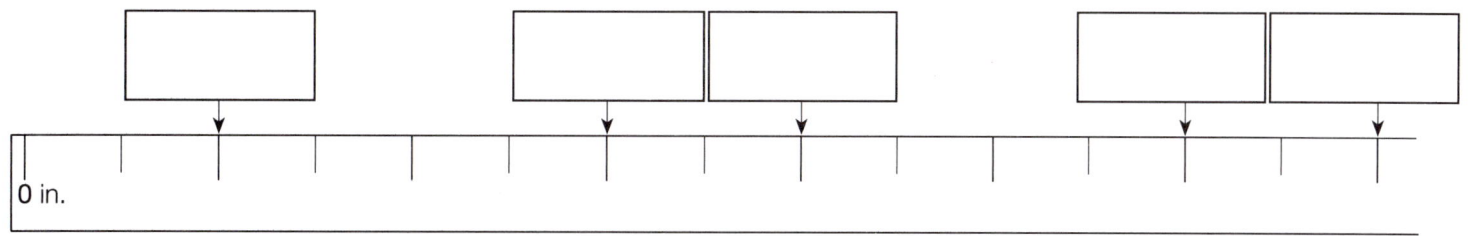

0 in.

3 Read the weight on each scale and write it below.

10 points per question

(1)

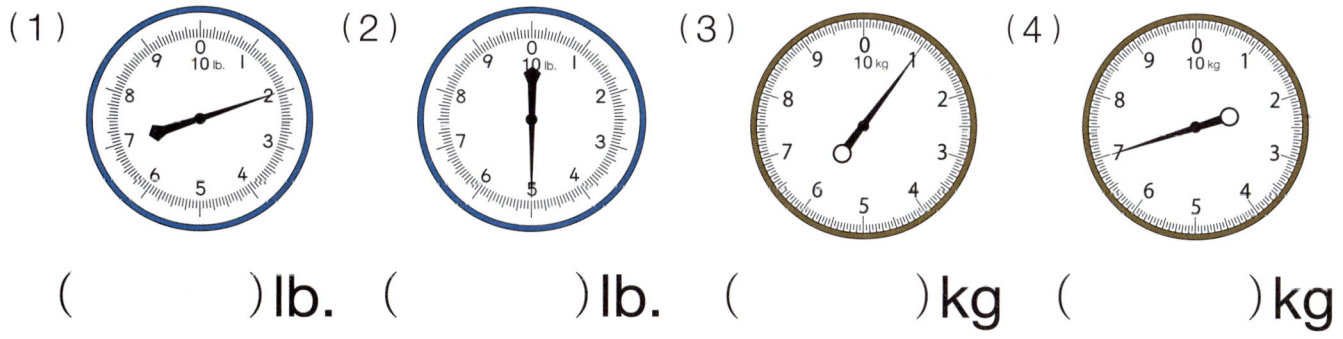

(   ) lb.

(2)

(   ) lb.

(3)

(   ) kg

(4)

(   ) kg

Reading

DAY

45

Review

Date / /

Name

Level ★★★

Score

/10

① Read the following book cover. Then answer the questions below using words from the cover.

15 points per question

Police Horses

by
Lee Camp

illustrated by
Sue Land

(1) What is the title, or name, of the book?
The title of the book is *Police* _____.

(2) Who is the author, or person who wrote the book?
The author is Lee _____ .

(3) Who is the illustrator, or person who drew the pictures in the book?
The illustrator is Sue _____.

② Read the passage. Then read the sentences below. Circle the "T" if the sentence is true, or correct. Circle the "F" if the sentence is false, or wrong.

11 points per question

Some police work with horses. Horses can take police to places where cars cannot go. Police and horses can save people on mountains. Horses are faster than people on foot. Horses can also be trained to find people. Horses can hear, smell and see very well.

(1) Horses can't help the police save people.   T   F

(2) Horses are faster than people on foot.   T   F

(3) Horses can be trained to find people.   T   F

(4) Horses can't hear very well.   T   F

(5) Police and horses can save people on mountains.   T   F

DAY 1, pages 1 & 2

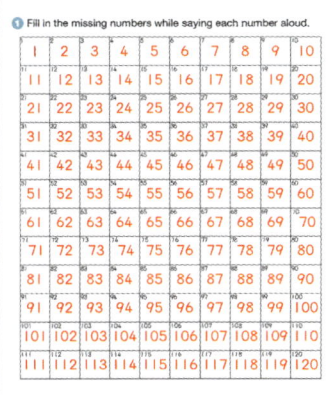

❶ Fill in the missing numbers while saying each number aloud.

1	2	3	4	5	6	7	8	9	10
11	12	13	14	15	16	17	18	19	20
21	22	23	24	25	26	27	28	29	30
31	32	33	34	35	36	37	38	39	40
41	42	43	44	45	46	47	48	49	50
51	52	53	54	55	56	57	58	59	60
61	62	63	64	65	66	67	68	69	70
71	72	73	74	75	76	77	78	79	80
81	82	83	84	85	86	87	88	89	90
91	92	93	94	95	96	97	98	99	100
101	102	103	104	105	106	107	108	109	110
111	112	113	114	115	116	117	118	119	120

❶ Trace the letters A to Z while saying each letter aloud.

A B C D / E F G H / I J K L / M N O P / Q R S T / U V W X / Y Z

DAY 2, pages 3 & 4

❶ Read each number sentence as you trace it.

(1) $1 + 1 = 2$ One plus one equals two
(2) $2 + 1 = 3$ Two plus one equals three
(3) $3 + 1 = 4$ Three plus one equals four
(4) $4 + 1 = 5$ Four plus one equals five
(5) $5 + 1 = 6$ Five plus one equals six
(6) $6 + 1 = 7$ Six plus one equals seven
(7) $7 + 1 = 8$ Seven plus one equals eight
(8) $8 + 1 = 9$ Eight plus one equals nine
(9) $9 + 1 = 10$ Nine plus one equals ten
(10) $10 + 1 = 11$ Ten plus one equals eleven

❷ Add.

(1) $3 + 1 = 4$ (6) $4 + 1 = 5$
(2) $6 + 1 = 7$ (7) $5 + 1 = 6$
(3) $7 + 1 = 8$ (8) $10 + 1 = 11$
(4) $2 + 1 = 3$ (9) $9 + 1 = 10$
(5) $1 + 1 = 2$ (10) $8 + 1 = 9$

❶ Trace the letters a to z while saying each letter aloud.

a b c d / e f g h / i j k l / m n o p / q r s t / u v w x / y z

DAY 3, pages 5 & 6

❶ Read each number sentence as you trace it and add. Use the examples as hints.

(1) $1 + 2 = 3$ One plus two equals three
(2) $2 + 2 = 4$ Two plus two equals four
(3) $3 + 2 = 5$ Three plus two equals five
(4) $4 + 2 = 6$ Four plus two equals six
(5) $5 + 2 = 7$ Five plus two equals seven
(6) $6 + 2 = 8$ Six plus two equals eight
(7) $7 + 2 = 9$ Seven plus two equals nine
(8) $8 + 2 = 10$
(9) $9 + 2 = 11$
(10) $10 + 2 = 12$

❷ Add.

(1) $3 + 2 = 5$ (6) $4 + 2 = 6$
(2) $6 + 2 = 8$ (7) $5 + 2 = 7$
(3) $7 + 2 = 9$ (8) $10 + 2 = 12$
(4) $1 + 2 = 3$ (9) $9 + 2 = 11$
(5) $2 + 2 = 4$ (10) $8 + 2 = 10$

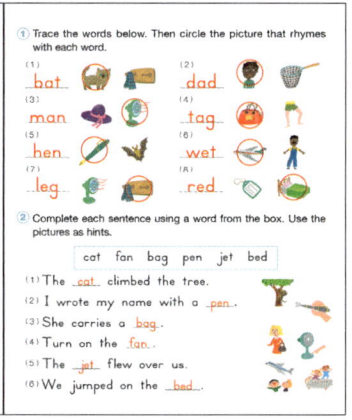

❶ Trace the words below. Then circle the picture that rhymes with each word.

(1) bat (2) dad (3) man (4) tag (5) hen (6) wet (7) leg (8) red

❷ Complete each sentence using a word from the box. Use the pictures as hints.

cat fan bag pen jet bed

(1) The cat climbed the tree.
(2) I wrote my name with a pen.
(3) She carries a bag.
(4) Turn on the fan.
(5) The jet flew over us.
(6) We jumped on the bed.

DAY 4, pages 7 & 8

❶ Read each number sentence as you trace it or say the number sentence aloud as you add. Use the examples as hints.

(1) $1 + 3 = 4$ One plus three equals four
(2) $2 + 3 = 5$ Two plus three equals five
(3) $3 + 3 = 6$ Three plus three equals six
(4) $4 + 3 = 7$ Four plus three equals seven
(5) $5 + 3 = 8$ Five plus three equals eight
(6) $6 + 3 = 9$
(7) $7 + 3 = 10$
(8) $8 + 3 = 11$
(9) $9 + 3 = 12$
(10) $10 + 3 = 13$

❷ Add.

(1) $3 + 3 = 6$ (6) $10 + 3 = 13$
(2) $1 + 3 = 4$ (7) $9 + 3 = 12$
(3) $2 + 3 = 5$ (8) $8 + 3 = 11$
(4) $4 + 3 = 7$ (9) $6 + 3 = 9$
(5) $5 + 3 = 8$ (10) $7 + 3 = 10$

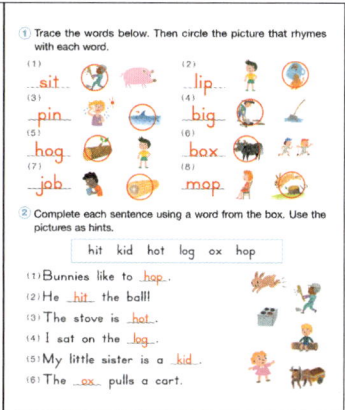

❶ Trace the words below. Then circle the picture that rhymes with each word.

(1) sit (2) lip (3) pin (4) big (5) hog (6) box (7) job (8) mop

❷ Complete each sentence using a word from the box. Use the pictures as hints.

hit kid hot log ox hop

(1) Bunnies like to hop.
(2) He hit the ball!
(3) The stove is hot.
(4) I sat on the log.
(5) My little sister is a kid.
(6) The ox pulls a cart.

DAY 5, pages 9 & 10

❶ Read each number sentence as you trace it or say the number sentence aloud as you add. Use the examples as hints.

(1) $1 + 4 = 5$ One plus four equals five
(2) $2 + 4 = 6$ Two plus four equals six
(3) $3 + 4 = 7$ Three plus four equals seven
(4) $4 + 4 = 8$
(5) $5 + 4 = 9$
(6) $6 + 4 = 10$
(7) $7 + 4 = 11$
(8) $8 + 4 = 12$
(9) $9 + 4 = 13$
(10) $10 + 4 = 14$

❷ Add.

(1) $3 + 4 = 7$ (6) $10 + 4 = 14$
(2) $1 + 4 = 5$ (7) $9 + 4 = 13$
(3) $2 + 4 = 6$ (8) $8 + 4 = 12$
(4) $4 + 4 = 8$ (9) $6 + 4 = 10$
(5) $5 + 4 = 9$ (10) $7 + 4 = 11$

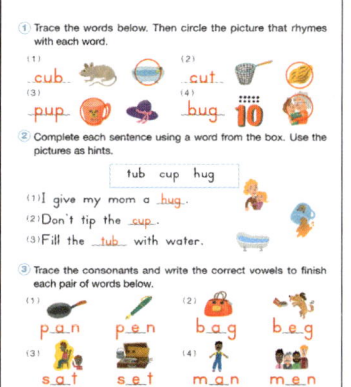

❶ Trace the words below. Then circle the picture that rhymes with each word.

(1) cub (2) cut (3) pup (4) bug

❷ Complete each sentence using a word from the box. Use the pictures as hints.

tub cup hug

(1) I give my mom a hug.
(2) Don't tip the cup.
(3) Fill the tub with water.

❸ Trace the consonants and write the correct vowels to finish each pair of words below.

(1) pan pen (2) bag beg (3) sat set (4) man men

DAY 6, pages 11 & 12

❶ Add. Use the examples as hints.

(1) $1 + 5 = 6$ (6) $6 + 5 = 11$
(2) $2 + 5 = 7$ (7) $7 + 5 = 12$
(3) $3 + 5 = 8$ (8) $8 + 5 = 13$
(4) $4 + 5 = 9$ (9) $9 + 5 = 14$
(5) $5 + 5 = 10$ (10) $10 + 5 = 15$

❷ Add.

(1) $1 + 6 = 7$ (6) $6 + 6 = 12$
(2) $2 + 6 = 8$ (7) $7 + 6 = 13$
(3) $3 + 6 = 9$ (8) $8 + 6 = 14$
(4) $4 + 6 = 10$ (9) $9 + 6 = 15$
(5) $5 + 6 = 11$ (10) $10 + 6 = 16$

❸ Add.

(1) $2 + 5 = 7$ (6) $7 + 5 = 12$
(2) $3 + 5 = 8$ (7) $8 + 5 = 13$
(3) $3 + 6 = 9$ (8) $9 + 6 = 15$
(4) $4 + 6 = 10$ (9) $10 + 6 = 16$
(5) $6 + 6 = 12$ (10) $9 + 5 = 14$

❶ Write the correct vowels to finish each pair of words below.

(1) pot pit (2) dog dig
(3) hot hit (4) rob rib
(5) cub cob (6) nut net
(7) bug big (8) hug hog

❷ Complete each sentence using a word from the box. Use the pictures as hints.

hot dig sun pan red

(1) The man added a pan.
(2) My bed is red.
(3) The pig likes to dig.
(4) The tea pot is hot.
(5) We have fun in the sun.

DAY 7, pages 13 & 14

❶ Add. Use the examples as hints.

(1) $1 + 7 = 8$ (6) $6 + 7 = 13$
(2) $2 + 7 = 9$ (7) $7 + 7 = 14$
(3) $3 + 7 = 10$ (8) $8 + 7 = 15$
(4) $4 + 7 = 11$ (9) $9 + 7 = 16$
(5) $5 + 7 = 12$ (10) $10 + 7 = 17$

❷ Add.

(1) $1 + 8 = 9$ (6) $6 + 8 = 14$
(2) $2 + 8 = 10$ (7) $7 + 8 = 15$
(3) $3 + 8 = 11$ (8) $8 + 8 = 16$
(4) $4 + 8 = 12$ (9) $9 + 8 = 17$
(5) $5 + 8 = 13$ (10) $10 + 8 = 18$

❸ Add.

(1) $3 + 7 = 10$ (6) $7 + 8 = 15$
(2) $4 + 7 = 11$ (7) $8 + 8 = 16$
(3) $3 + 8 = 11$ (8) $9 + 7 = 16$
(4) $3 + 8 = 11$ (9) $7 + 7 = 14$
(5) $1 + 7 = 8$ (10) $10 + 8 = 18$

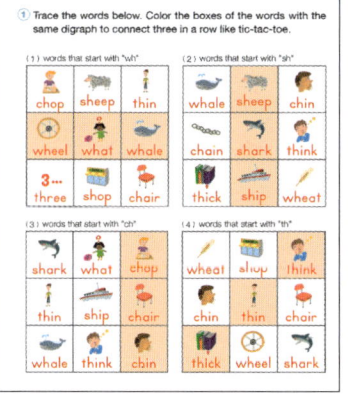

❶ Trace the words below. Color the boxes of the words with the same digraph to connect three in a row like tic-tac-toe.

(1) words that start with "wh"

chop	sheep	thin
wheel	what	whale
three	shop	chair

(2) words that start with "sh"

whale	sheep	chin
chain	shark	think
thick	ship	wheat

(3) words that start with "ch"

shark	what	chop
thin	ship	chair
whale	think	chin

(4) words that start with "th"

wheat	ship	think
chin	thin	chair
thick	wheel	shark

DAY 8, pages 15 & 16

❶ Add. Use the example as a hint.

(1) $1 + 9 = 10$ (6) $6 + 9 = 15$
(2) $2 + 9 = 11$ (7) $7 + 9 = 16$
(3) $3 + 9 = 12$ (8) $8 + 9 = 17$
(4) $4 + 9 = 13$ (9) $9 + 9 = 18$
(5) $5 + 9 = 14$ (10) $10 + 9 = 19$

❷ Add. Use the example as a hint.

(1) $1 + 10 = 11$ (6) $6 + 10 = 16$
(2) $2 + 10 = 12$ (7) $7 + 10 = 17$
(3) $3 + 10 = 13$ (8) $8 + 10 = 18$
(4) $4 + 10 = 14$ (9) $9 + 10 = 19$
(5) $5 + 10 = 15$ (10) $10 + 10 = 20$

❸ Add.

(1) $1 + 9 = 10$ (6) $9 + 10 = 19$
(2) $2 + 9 = 11$ (7) $10 + 10 = 20$
(3) $4 + 10 = 14$ (8) $9 + 9 = 18$
(4) $5 + 10 = 15$ (9) $8 + 9 = 17$
(5) $8 + 10 = 18$ (10) $4 + 9 = 13$

❶ Trace the words below. Color the boxes of the words with the same consonant blend to connect three in a row like tic-tac-toe.

(1) words that start with "bl"

block	float	clock
flag	blast	plum
plant	climb	block

(2) words that start with "cl"

block	clap	plane
fly	clean	flat
blimp	claw	plant

(3) words that start with "fl"

clock	plane	float
blimp	flute	blast
flap	plate	clue

(4) words that start with "pl"

play	plow	plant
clap	flap	blue
block	fly	clam

DAY 9, pages 17 & 18

① Add.
(1) 7 + 1 = 8
(2) 9 + 1 = 10
(3) 6 + 1 = 7
(4) 6 + 2 = 8
(5) 8 + 2 = 10
(6) 5 + 2 = 7
(7) 6 + 3 = 9
(8) 5 + 3 = 8
(9) 1 + 4 = 5
(10) 3 + 4 = 7
(11) 4 + 5 = 9
(12) 7 + 5 = 12

② Add.
(1) 1 + 6 = 7
(2) 2 + 6 = 8
(3) 3 + 6 = 9
(4) 4 + 7 = 11
(5) 5 + 7 = 12
(6) 3 + 7 = 10
(7) 4 + 8 = 12
(8) 5 + 8 = 13
(9) 6 + 8 = 14
(10) 6 + 9 = 15
(11) 5 + 9 = 14
(12) 2 + 9 = 11
(13) 3 + 9 = 12
(14) 3 + 10 = 13
(15) 5 + 10 = 15
(16) 7 + 10 = 17

① Trace the words below. Color the boxes of the words with the same consonant blend to connect three in a row like tic-tac-toe.

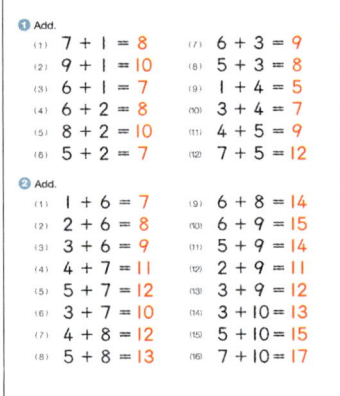

(1) words that start with "br": crib, frog, fry / grass, gray, crack / brick, brown, brain
(2) words that start with "cr": green, fry, crab / brain, grin, cry / fruit, brick, crack
(3) words that start with "fr": frog, grape, crawl / grab, fry, braid / cry, brag, fruit
(4) words that start with "gr": grape, brown, crib / grass, fry, brick / green, crawl, fruit

DAY 10, pages 19 & 20

① Add.
(1) 5 + 1 = 6
(2) 5 + 2 = 7
(3) 5 + 3 = 8
(4) 4 + 5 = 9
(5) 4 + 6 = 10
(6) 4 + 7 = 11
(7) 8 + 3 = 11
(8) 8 + 4 = 12
(9) 8 + 5 = 13
(10) 7 + 7 = 14
(11) 7 + 8 = 15
(12) 7 + 9 = 16

② Add.
(1) 7 + 4 = 11
(2) 8 + 3 = 11
(3) 6 + 5 = 11
(4) 9 + 6 = 15
(5) 5 + 8 = 13
(6) 2 + 7 = 9
(7) 5 + 10 = 15
(8) 3 + 9 = 12
(9) 5 + 6 = 11
(10) 2 + 4 = 6
(11) 9 + 4 = 13
(12) 7 + 1 = 8
(13) 9 + 9 = 18
(14) 8 + 5 = 13
(15) 3 + 10 = 13
(16) 4 + 5 = 9

① Trace the words below. Color the boxes of the words with the same consonant blend to connect three in a row like tic-tac-toe.

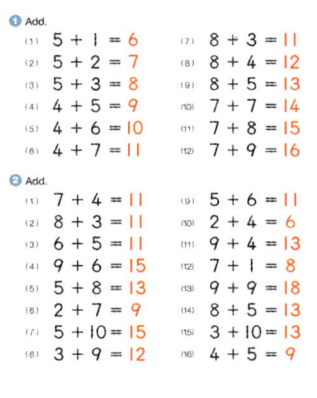

(1) words that start with "sk": smart, smoke, space / sky, skip, skirt / spy, snow, snake
(2) words that start with "sm": ski, smell, skip / snail, small, snow / spot, smile, speak
(3) words that start with "sn": spoon, smoke, snake / small, ski, snow / space, skin, snap
(4) words that start with "sp": snap, smell, spy / sky, spot, skirt / spoon, smart, snow

DAY 11, pages 21 & 22

① Add.
(1) 9 + 1 = 10
(2) 10 + 1 = 11
(3) 11 + 1 = 12
(4) 10 + 2 = 12
(5) 11 + 2 = 13
(6) 12 + 2 = 14
(7) 12 + 3 = 15
(8) 13 + 3 = 16
(9) 14 + 3 = 17
(10) 12 + 4 = 16
(11) 13 + 4 = 17
(12) 14 + 4 = 18

② Add.
(1) 12 + 5 = 17
(2) 13 + 5 = 18
(3) 14 + 5 = 19
(4) 14 + 6 = 20
(5) 12 + 6 = 18
(6) 13 + 6 = 19
(7) 13 + 7 = 20
(8) 12 + 7 = 19
(9) 11 + 7 = 18
(10) 11 + 8 = 19
(11) 12 + 8 = 20
(12) 10 + 8 = 18
(13) 10 + 9 = 19
(14) 11 + 9 = 20
(15) 15 + 5 = 20
(16) 11 + 6 = 17

① Trace the words below. Color the boxes of the words with the same consonant blend to connect three in a row like tic-tac-toe.

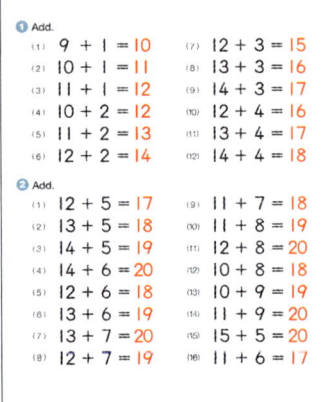

(1) words that start with "gl": truck, start, slip / glad, glove, glow / trip, slow, stew
(2) words that start with "sl": train, slow, glad / step, slip, trail / stamp, sleep, glass
(3) words that start with "st": slow, slug, start / trash, glad, step / tree, glue, stamp
(4) words that start with "tr": step, glow, truck / stew, train, slow / tree, sleep, glove

DAY 12, pages 23 & 24

① Add.
(1) 12 + 2 = 14
(2) 14 + 2 = 16
(3) 16 + 2 = 18
(4) 12 + 3 = 15
(5) 14 + 3 = 17
(6) 16 + 3 = 19
(7) 12 + 4 = 16
(8) 14 + 4 = 18
(9) 16 + 4 = 20
(10) 12 + 5 = 17
(11) 14 + 5 = 19
(12) 16 + 5 = 21

② Add.
(1) 12 + 6 = 18
(2) 13 + 6 = 19
(3) 14 + 6 = 20
(4) 14 + 7 = 21
(5) 12 + 7 = 19
(6) 13 + 8 = 21
(7) 14 + 8 = 22
(8) 12 + 9 = 21
(9) 17 + 4 = 21
(10) 17 + 5 = 22
(11) 17 + 7 = 24
(12) 18 + 3 = 21
(13) 18 + 5 = 23
(14) 18 + 6 = 24
(15) 19 + 3 = 22
(16) 19 + 5 = 24

If you want more addition practice, check out Kumon's Addition Grade 1

① Complete each sentence by using words from the box below.

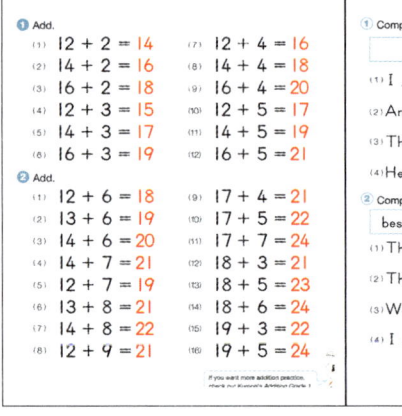

box: elephant jump best ink
(1) I jump high.
(2) An elephant has large ears.
(3) There is ink on the desk.
(4) He was the best runner.

② Complete each phrase by using words from the box below.

box: best ant tank stamp tent test lamp sink
(1) The ant went into the tent.
(2) The stamp shows a lamp.
(3) We saw the tank sink.
(4) I did my best on the test.

DAY 13, pages 25 & 26

① Add.
(1) 3 + 2 = 5
(6) 3 + 2 = 5 (With the answer here)
(11) 6 + 3 = 9
(16) 6 + 3 = 9
(2) 5 + 4 = 9
(7) 5 + 4 = 9
(12) 8 + 1 = 9
(17) 8 + 1 = 9
(3) 7 + 1 = 8
(8) 7 + 1 = 8
(13) 5 + 3 = 8
(18) 5 + 3 = 8
(4) 3 + 4 = 7
(9) 3 + 4 = 7
(14) 6 + 1 = 7
(19) 6 + 1 = 7
(5) 2 + 6 = 8
(10) 2 + 6 = 8
(15) 4 + 3 = 7
(20) 4 + 3 = 7

① Complete each sentence by using words from the box below.

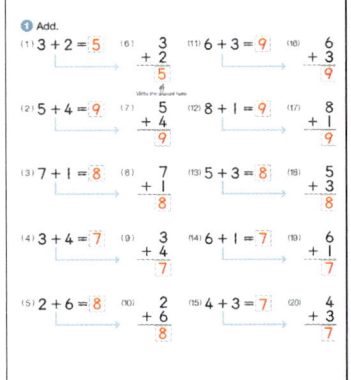

box: fish lunch truck sand
(1) The truck is here.
(2) I put my hand in the sand.
(3) The fish swam away.
(4) She sat to eat her lunch.

② Complete each phrase by using words from the box below.

box: wash bench wand rock lunch duck dish hand
(1) I wash the dish.
(2) We ate lunch on the bench.
(3) I held the wand in my hand.
(4) The duck sat on a rock.

DAY 14, pages 27 & 28

① Say the number sentence aloud as you subtract. Use the examples as hints.
(1) 2 − 1 = 1 (Two minus one equals one)
(2) 3 − 1 = 2 (Three minus one equals two)
(3) 4 − 1 = 3
(4) 5 − 1 = 4
(5) 6 − 1 = 5
(6) 7 − 1 = 6
(7) 8 − 1 = 7
(8) 9 − 1 = 8
(9) 10 − 1 = 9
(10) 11 − 1 = 10

② Subtract.
(1) 2 − 1 = 1
(2) 3 − 1 = 2
(3) 5 − 1 = 4
(4) 6 − 1 = 5
(6) 7 − 1 = 6
(7) 9 − 1 = 8
(8) 8 − 1 = 7
(9) 10 − 1 = 9
(10) 11 − 1 = 10

1 2 3 4 5 6 7 8 9 10 11

① Complete each sentence by using words from the box below.

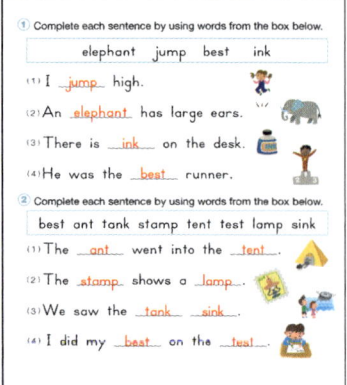

box: sniff summer carrot class funny ill
(1) I felt ill.
(2) He laughed at the funny joke.
(3) The carrot is orange.
(4) We sat in the class.
(5) The boy took a sniff.
(6) In the summer, we sail boats.

② Complete each phrase by using words from the box below.

box: offer bull sunny off summer grass tall berry
(1) I will offer some berry juice.
(2) We walked in the tall grass.
(3) It's sunny in the summer.
(4) The bull threw the cowboy off his bac[k]

DAY 15, pages 29 & 30

① Say the number sentence aloud as you subtract. Use the examples as hints.
(1) 3 − 2 = 1
(2) 4 − 2 = 2
(3) 5 − 2 = 3
(4) 6 − 2 = 4
(5) 7 − 2 = 5
(6) 8 − 2 = 6
(7) 9 − 2 = 7
(8) 10 − 2 = 8
(9) 11 − 2 = 9
(10) 12 − 2 = 10

② Subtract.
(1) 3 − 2 = 1
(2) 4 − 2 = 2
(3) 9 − 2 = 7
(4) 7 − 2 = 5
(5) 8 − 2 = 6
(6) 10 − 2 = 8
(7) 5 − 2 = 3
(8) 6 − 2 = 4
(9) 12 − 2 = 10
(10) 11 − 2 = 9

1 2 3 4 5 6 7 8 9 10 11 12

① Pick the correct word from the box to match each picture below.

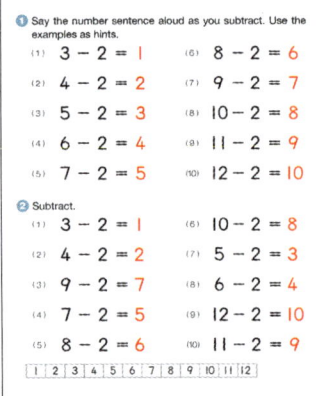

box: play bee mane read rain meet
(1) play (2) meet (3) rain
(4) bee (5) read (6) mane

② Fill in the missing vowels. Use the pictures as clues.
(1) mane (2) meet
(3) read (4) play
(5) rain (6) bee

③ Trace the words below. Circle the word with the matching long vowel sound.
(1) Long "a" as in plane: (mane) seat fan
(2) Long "e" as in tree: sky (bee) net

DAY 16, pages 31 & 32

① Say the number sentence aloud as you subtract. Use the examples as hints.
(1) 4 − 3 = 1
(2) 5 − 3 = 2
(3) 6 − 3 = 3
(4) 7 − 3 = 4
(5) 8 − 3 = 5
(6) 9 − 3 = 6
(7) 10 − 3 = 7
(8) 11 − 3 = 8
(9) 12 − 3 = 9
(10) 13 − 3 = 10

② Subtract.
(1) 4 − 3 = 1
(2) 5 − 3 = 2
(3) 10 − 3 = 7
(4) 11 − 3 = 8
(5) 12 − 3 = 9
(6) 8 − 3 = 5
(7) 9 − 3 = 6
(8) 6 − 3 = 3
(9) 7 − 3 = 4
(10) 13 − 3 = 10

1 2 3 4 5 6 7 8 9 10 11 12 13

① Pick the correct word from the box to match each picture below.

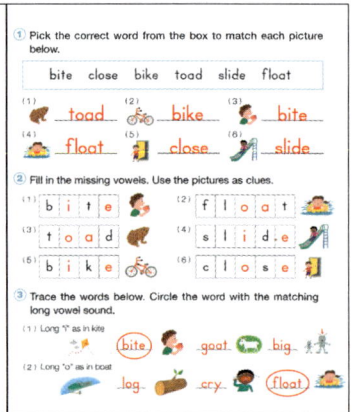

box: bite close bike toad slide float
(1) toad (2) bike (3) bite
(4) float (5) close (6) slide

② Fill in the missing vowels. Use the pictures as clues.
(1) bite (2) float
(3) toad (4) slide
(5) bike (6) close

③ Trace the words below. Circle the word with the matching long vowel sound.
(1) Long "i" as in kite: (bite) goat big
(2) Long "o" as in boat: log cry (float)

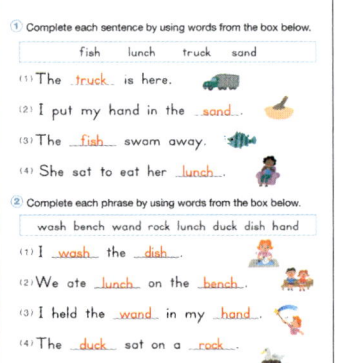
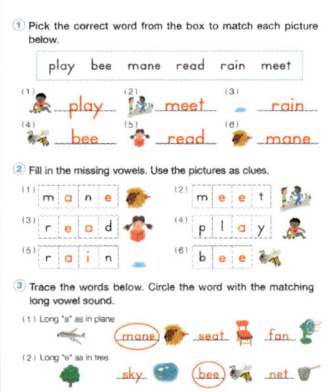
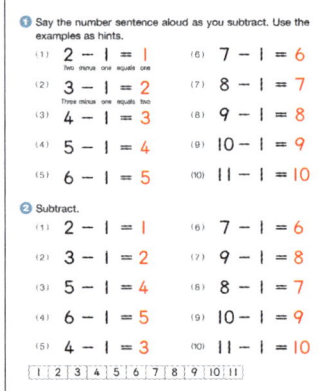
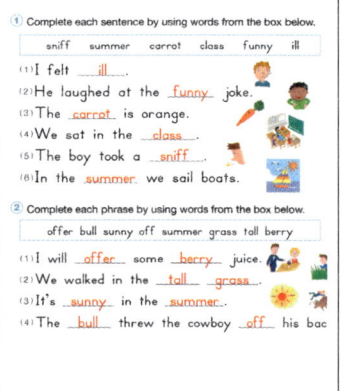
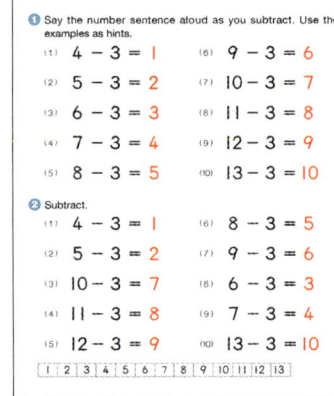

DAY 17, pages 33 & 34

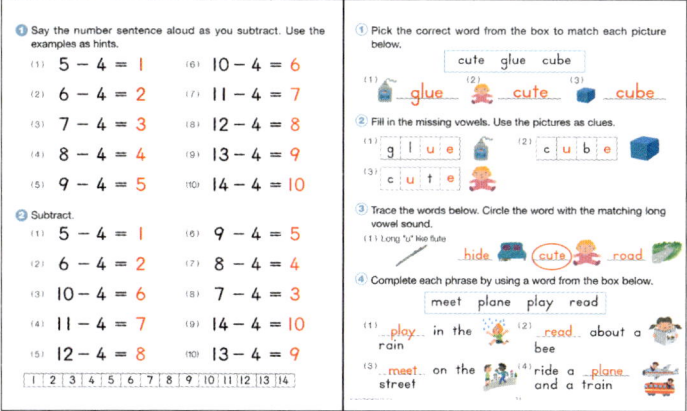

① Say the number sentence aloud as you subtract. Use the examples as hints.

(1) 5 − 4 = 1   (6) 10 − 4 = 6
(2) 6 − 4 = 2   (7) 11 − 4 = 7
(3) 7 − 4 = 3   (8) 12 − 4 = 8
(4) 8 − 4 = 4   (9) 13 − 4 = 9
(5) 9 − 4 = 5   (10) 14 − 4 = 10

② Subtract.

(1) 5 − 4 = 1   (6) 9 − 4 = 5
(2) 6 − 4 = 2   (7) 8 − 4 = 4
(3) 10 − 4 = 6   (8) 7 − 4 = 3
(4) 11 − 4 = 7   (9) 14 − 4 = 10
(5) 12 − 4 = 8   (10) 13 − 4 = 9

| 1 | 2 | 3 | 4 | 5 | 6 | 7 | 8 | 9 | 10 | 11 | 12 | 13 | 14 |

① Pick the correct word from the box to match each picture below.

cute glue cube

(1) glue   (2) cute   (3) cube

② Fill in the missing vowels. Use the pictures as clues.

(1) g l u e   (2) c u b e
(3) c u t e

③ Trace the words below. Circle the word with the matching long vowel sound.
(1) Long "u" like flute

hide   (cute)   road

④ Complete each phrase by using a word from the box below.

meet plane play read

(1) play in the rain   (2) read about a bee
(3) meet on the street   (4) ride a plane and a train

DAY 18, pages 35 & 36

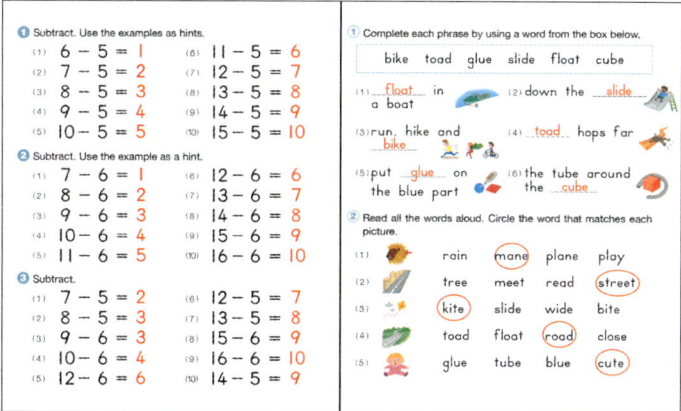

① Subtract. Use the examples as hints.

(1) 6 − 5 = 1   (6) 11 − 5 = 6
(2) 7 − 5 = 2   (7) 12 − 5 = 7
(3) 8 − 5 = 3   (8) 13 − 5 = 8
(4) 9 − 5 = 4   (9) 14 − 5 = 9
(5) 10 − 5 = 5   (10) 15 − 5 = 10

② Subtract. Use the example as a hint.

(1) 7 − 6 = 1   (6) 12 − 6 = 6
(2) 8 − 6 = 2   (7) 13 − 6 = 7
(3) 9 − 6 = 3   (8) 14 − 6 = 8
(4) 10 − 6 = 4   (9) 15 − 6 = 9
(5) 11 − 6 = 5   (10) 16 − 6 = 10

③ Subtract.

(1) 7 − 5 = 2   (6) 12 − 5 = 7
(2) 8 − 5 = 3   (7) 13 − 5 = 8
(3) 9 − 5 = 3   (8) 15 − 6 = 9
(4) 10 − 5 = 4   (9) 16 − 6 = 10
(5) 12 − 6 = 6   (10) 14 − 5 = 9

① Complete each phrase by using a word from the box below.

bike toad glue slide float cube

(1) float in a boat   (2) down the slide
(3) run, hike and bike   (4) toad hops far
(5) put glue on the blue part   (6) the tube around the cube

② Read all the words aloud. Circle the word that matches each picture.

(1) rain   (mane)   plane   play
(2) tree   meet   read   (street)
(3) (kite)   slide   wide   bite
(4) toad   float   (road)   close
(5) glue   tube   blue   (cute)

DAY 19, pages 37 & 38

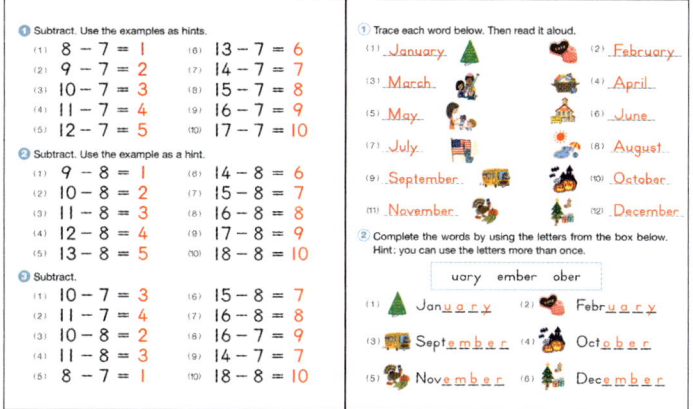

① Subtract. Use the examples as hints.

(1) 8 − 7 = 1   (6) 13 − 7 = 6
(2) 9 − 7 = 2   (7) 14 − 7 = 7
(3) 10 − 7 = 3   (8) 15 − 7 = 8
(4) 11 − 7 = 4   (9) 16 − 7 = 9
(5) 12 − 7 = 5   (10) 17 − 7 = 10

② Subtract. Use the example as a hint.

(1) 9 − 8 = 1   (6) 14 − 8 = 6
(2) 10 − 8 = 2   (7) 15 − 8 = 7
(3) 11 − 8 = 3   (8) 16 − 8 = 8
(4) 12 − 8 = 4   (9) 17 − 8 = 9
(5) 13 − 8 = 5   (10) 18 − 8 = 10

③ Subtract.

(1) 10 − 7 = 3   (6) 15 − 8 = 7
(2) 11 − 7 = 4   (7) 16 − 8 = 8
(3) 10 − 8 = 2   (8) 16 − 7 = 9
(4) 11 − 8 = 3   (9) 14 − 7 = 7
(5) 8 − 7 = 1   (10) 18 − 8 = 10

① Trace each word below. Then read it aloud.

(1) January   (2) February
(3) March   (4) April
(5) May   (6) June
(7) July   (8) August
(9) September   (10) October
(11) November   (12) December

② Complete the words by using the letters from the box below. Hint: you can use the letters more than once.

uary ember ober

(1) January   (2) February
(3) September   (4) October
(5) November   (6) December

DAY 20, pages 39 & 40

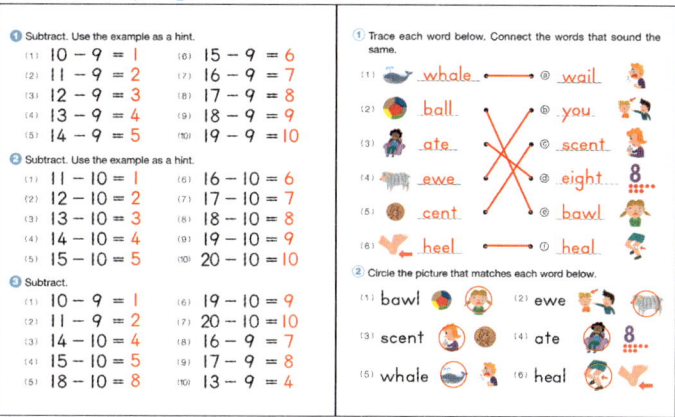

① Subtract. Use the example as a hint.

(1) 10 − 9 = 1   (6) 15 − 9 = 6
(2) 11 − 9 = 2   (7) 16 − 9 = 7
(3) 12 − 9 = 3   (8) 17 − 9 = 8
(4) 13 − 9 = 4   (9) 18 − 9 = 9
(5) 14 − 9 = 5   (10) 19 − 9 = 10

② Subtract. Use the example as a hint.

(1) 11 − 10 = 1   (6) 16 − 10 = 6
(2) 12 − 10 = 2   (7) 17 − 10 = 7
(3) 13 − 10 = 3   (8) 18 − 10 = 8
(4) 14 − 10 = 4   (9) 19 − 10 = 9
(5) 15 − 10 = 5   (10) 20 − 10 = 10

③ Subtract.

(1) 10 − 9 = 1   (6) 19 − 10 = 9
(2) 11 − 9 = 2   (7) 20 − 10 = 10
(3) 14 − 10 = 4   (8) 16 − 9 = 7
(4) 15 − 10 = 5   (9) 17 − 9 = 8
(5) 18 − 10 = 8   (10) 13 − 9 = 4

① Trace each word below. Connect the words that sound the same.

(1) whale — (a) wail
(2) ball — (b) you
(3) ate — (c) scent
(4) ewe — (d) eight
(5) cent — (e) bawl
(6) heel — (f) heal

② Circle the picture that matches each word below.

(1) bawl   (2) ewe
(3) scent   (4) ate
(5) whale   (6) heal

DAY 21, pages 41 & 42

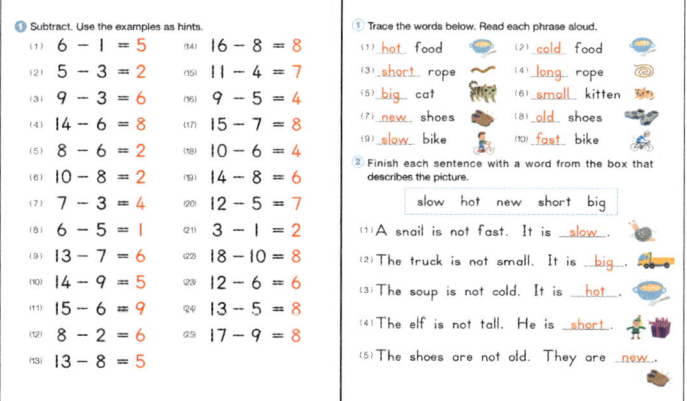

① Subtract. Use the examples as hints.

(1) 6 − 1 = 5   (14) 16 − 8 = 8
(2) 5 − 3 = 2   (15) 11 − 4 = 7
(3) 9 − 3 = 6   (16) 9 − 5 = 4
(4) 14 − 6 = 8   (17) 15 − 7 = 8
(5) 8 − 6 = 2   (18) 10 − 6 = 4
(6) 10 − 8 = 2   (19) 14 − 8 = 6
(7) 7 − 3 = 4   (20) 12 − 5 = 7
(8) 6 − 5 = 1   (21) 3 − 1 = 2
(9) 13 − 7 = 6   (22) 18 − 10 = 8
(10) 14 − 9 = 5   (23) 12 − 6 = 6
(11) 15 − 6 = 9   (24) 13 − 5 = 8
(12) 8 − 2 = 6   (25) 17 − 9 = 8
(13) 13 − 8 = 5

① Trace the words below. Read each phrase aloud.

(1) hot food   (2) cold food
(3) short rope   (4) long rope
(5) big cat   (6) small kitten
(7) new shoes   (8) old shoes
(9) slow bike   (10) fast bike

② Finish each sentence with a word from the box that describes the picture.

slow hot new short big

(1) A snail is not fast. It is slow.
(2) The truck is not small. It is big.
(3) The soup is not cold. It is hot.
(4) The elf is not tall. He is short.
(5) The shoes are not old. They are new.

DAY 22, pages 43 & 44

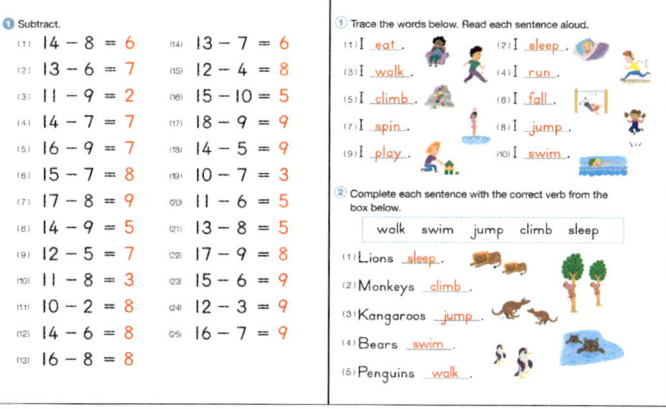

① Subtract.

(1) 14 − 8 = 6   (14) 13 − 7 = 6
(2) 13 − 6 = 7   (15) 12 − 4 = 8
(3) 11 − 9 = 2   (16) 15 − 10 = 5
(4) 14 − 7 = 7   (17) 18 − 9 = 9
(5) 16 − 9 = 7   (18) 14 − 5 = 9
(6) 15 − 7 = 8   (19) 10 − 7 = 3
(7) 17 − 8 = 9   (20) 11 − 6 = 5
(8) 14 − 9 = 5   (21) 13 − 8 = 5
(9) 12 − 5 = 7   (22) 17 − 9 = 8
(10) 11 − 8 = 3   (23) 15 − 6 = 9
(11) 10 − 2 = 8   (24) 12 − 3 = 9
(12) 14 − 6 = 8   (25) 16 − 7 = 9
(13) 16 − 8 = 8

① Trace the words below. Read each sentence aloud.

(1) I eat.   (2) I sleep.
(3) I walk.   (4) I run.
(5) I climb.   (6) I fall.
(7) I spin.   (8) I jump.
(9) I play.   (10) I swim.

② Complete each sentence with the correct verb from the box below.

walk swim jump climb sleep

(1) Lions sleep.
(2) Monkeys climb.
(3) Kangaroos jump.
(4) Bears swim.
(5) Penguins walk.

DAY 23, pages 45 & 46

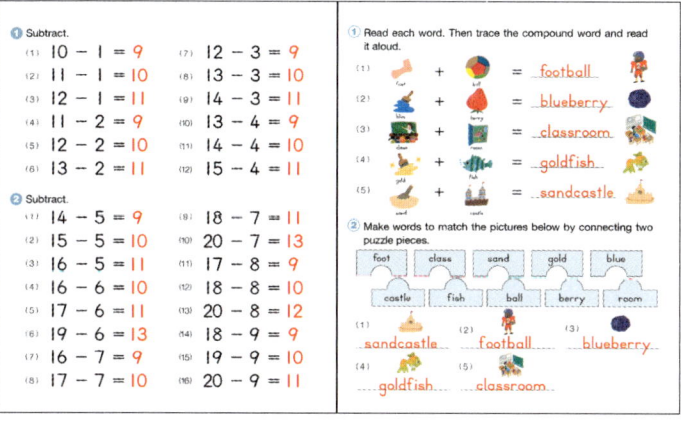

① Subtract.

(1) 10 − 1 = 9   (7) 12 − 3 = 9
(2) 11 − 1 = 10   (8) 13 − 3 = 10
(3) 12 − 1 = 11   (9) 14 − 3 = 11
(4) 11 − 2 = 9   (10) 13 − 4 = 9
(5) 12 − 2 = 10   (11) 14 − 4 = 10
(6) 13 − 2 = 11   (12) 15 − 4 = 11

② Subtract.

(1) 14 − 5 = 9   (9) 18 − 7 = 11
(2) 15 − 5 = 10   (10) 20 − 7 = 13
(3) 16 − 5 = 11   (11) 17 − 8 = 9
(4) 16 − 6 = 10   (12) 18 − 8 = 10
(5) 17 − 6 = 11   (13) 20 − 8 = 12
(6) 19 − 6 = 13   (14) 18 − 9 = 9
(7) 16 − 7 = 9   (15) 19 − 9 = 10
(8) 17 − 7 = 10   (16) 20 − 9 = 11

① Read each word. Then trace the compound word and read it aloud.

(1) + = football
(2) + = blueberry
(3) + = classroom
(4) + = goldfish
(5) + = sandcastle

② Make words to match the pictures below by connecting two puzzle pieces.

foot class sand gold blue
castle fish ball berry room

(1) sandcastle   (2) football   (3) blueberry
(4) goldfish   (5) classroom

DAY 24, pages 47 & 48

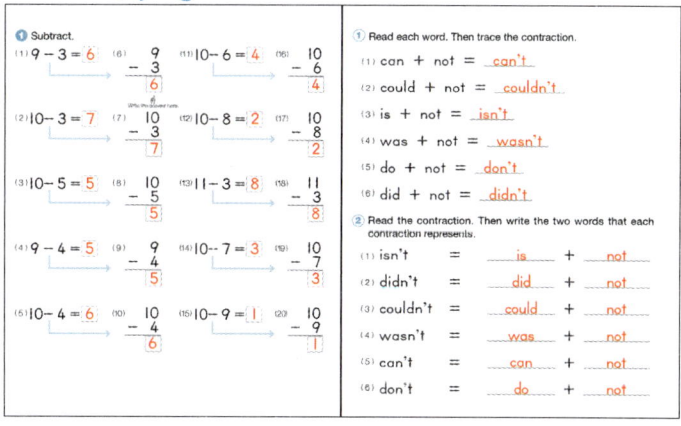

① Subtract.

(1) 9 − 3 = 6
(6) 9 − 3 = 6
(11) 10 − 6 = 4
(16) 10 − 6 = 4

(2) 10 − 3 = 7
(7) 10 − 3 = 7
(12) 10 − 8 = 2
(17) 10 − 8 = 2

(3) 10 − 5 = 5
(8) 10 − 5 = 5
(13) 11 − 3 = 8
(18) 11 − 3 = 8

(4) 9 − 4 = 5
(9) 9 − 4 = 5
(14) 10 − 7 = 3
(19) 10 − 7 = 3

(5) 10 − 4 = 6
(10) 10 − 4 = 6
(15) 10 − 9 = 1
(20) 10 − 9 = 1

① Read each word. Then trace the contraction.

(1) can + not = can't
(2) could + not = couldn't
(3) is + not = isn't
(4) was + not = wasn't
(5) do + not = don't
(6) did + not = didn't

② Read the contraction. Then write the two words that each contraction represents.

(1) isn't = is + not
(2) didn't = did + not
(3) couldn't = could + not
(4) wasn't = was + not
(5) can't = can + not
(6) don't = do + not

DAY 25, pages 49 & 50

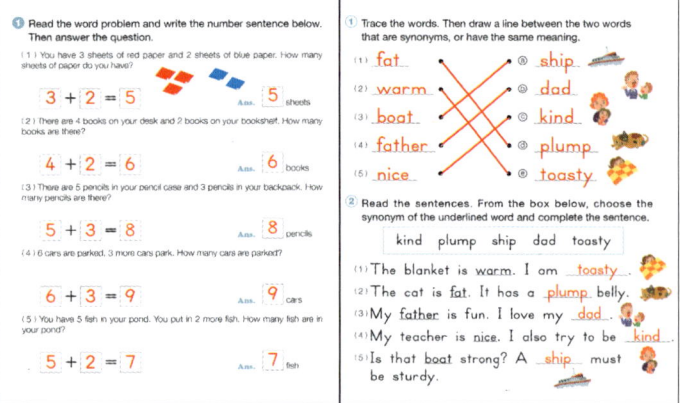

① Read the word problem and write the number sentence below. Then answer the question.

(1) You have 3 sheets of red paper and 2 sheets of blue paper. How many sheets of paper do you have?
$3 + 2 = 5$ Ans. 5 sheets

(2) There are 4 books on your desk and 2 books on your bookshelf. How many books are there?
$4 + 2 = 6$ Ans. 6 books

(3) There are 5 pencils in your pencil case and 3 pencils in your backpack. How many pencils are there?
$5 + 3 = 8$ Ans. 8 pencils

(4) 6 cars are parked. 3 more cars park. How many cars are parked?
$6 + 3 = 9$ Ans. 9 cars

(5) You have 5 fish in your pond. You put in 2 more fish. How many fish are in your pond?
$5 + 2 = 7$ Ans. 7 fish

① Trace the words. Then draw a line between the two words that are synonyms, or have the same meaning.
(1) fat — ⓔ plump
(2) warm — ⓔ toasty
(3) boat — ⓐ ship
(4) father — ⓑ dad
(5) nice — ⓒ kind

② Read the sentences. From the box below, choose the synonym of the underlined word and complete the sentence.
kind plump ship dad toasty
(1) The blanket is warm. I am toasty.
(2) The cat is fat. It has a plump belly.
(3) My father is fun. I love my dad.
(4) My teacher is nice. I also try to be kind.
(5) Is that boat strong? A ship must be sturdy.

DAY 26, pages 51 & 52

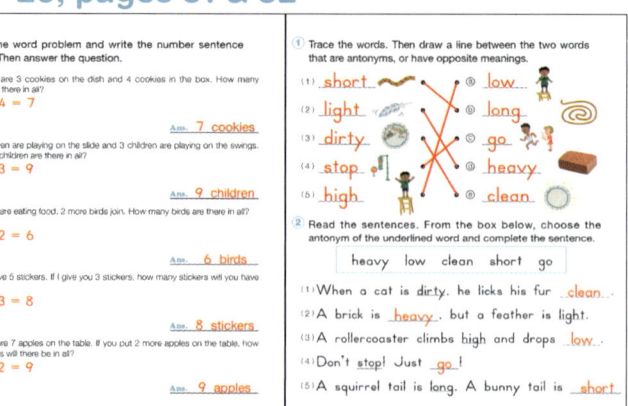

① Read the word problem and write the number sentence below. Then answer the question.

(1) There are 3 cookies on the dish and 4 cookies in the box. How many cookies are there in all?
$3 + 4 = 7$ Ans. 7 cookies

(2) 6 children are playing on the slide and 3 children are playing on the swings. How many children are there in all?
$6 + 3 = 9$ Ans. 9 children

(3) 4 birds are eating food. 2 more birds join. How many birds are there in all?
$4 + 2 = 6$ Ans. 6 birds

(4) You have 5 stickers. If I give you 3 stickers, how many stickers will you have in all?
$5 + 3 = 8$ Ans. 8 stickers

(5) There are 7 apples on the table. If you put 2 more apples on the table, how many apples will there be in all?
$7 + 2 = 9$ Ans. 9 apples

① Trace the words. Then draw a line between the two words that are antonyms, or have opposite meanings.
(1) short — ⓑ long
(2) light — ⓔ heavy
(3) dirty — ⓐ clean
(4) stop — ⓒ go
(5) high — ⓓ low

② Read the sentences. From the box below, choose the antonym of the underlined word and complete the sentence.
heavy low clean short go
(1) When a cat is dirty, he licks his fur clean.
(2) A brick is heavy, but a feather is light.
(3) A rollercoaster climbs high and drops low.
(4) Don't stop! Just go!
(5) A squirrel tail is long. A bunny tail is short.

DAY 27, pages 53 & 54

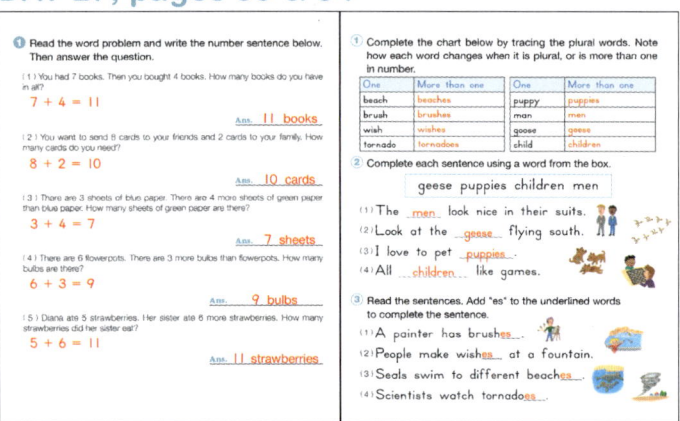

① Read the word problem and write the number sentence below. Then answer the question.

(1) You had 7 books. Then you bought 4 books. How many books do you have in all?
$7 + 4 = 11$ Ans. 11 books

(2) You want to send 8 cards to your friends and 2 cards to your family. How many cards do you need?
$8 + 2 = 10$ Ans. 10 cards

(3) There are 3 sheets of blue paper. There are 4 more sheets of green paper than blue paper. How many sheets of green paper are there?
$3 + 4 = 7$ Ans. 7 sheets

(4) There are 6 flowerpots. There are 3 more bulbs than flowerpots. How many bulbs are there?
$6 + 3 = 9$ Ans. 9 bulbs

(5) Diana ate 5 strawberries. Her sister ate 6 more strawberries. How many strawberries did her sister eat?
$5 + 6 = 11$ Ans. 11 strawberries

① Complete the chart below by tracing the plural words. Note how each word changes when it is plural, or is more than one in number.

One	More than one	One	More than one
beach	beaches	puppy	puppies
brush	brushes	man	men
wish	wishes	goose	geese
tornado	tornadoes	child	children

② Complete each sentence using a word from the box.
geese puppies children men
(1) The men look nice in their suits.
(2) Look at the geese flying south.
(3) I love to pet puppies.
(4) All children like games.

③ Read the sentences. Add "es" to the underlined words to complete the sentence.
(1) A painter has brushes.
(2) People make wishes at a fountain.
(3) Seals swim to different beaches.
(4) Scientists watch tornadoes.

DAY 28, pages 55 & 56

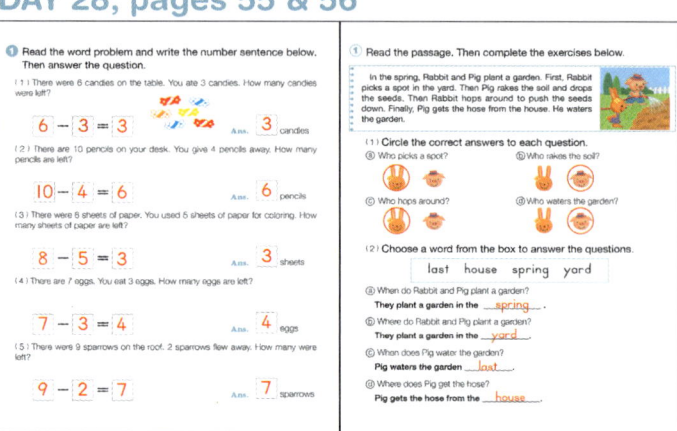

① Read the word problem and write the number sentence below. Then answer the question.

(1) There were 6 candies on the table. You ate 3 candies. How many candies were left?
$6 - 3 = 3$ Ans. 3 candies

(2) There are 10 pencils on your desk. You give 4 pencils away. How many pencils are left?
$10 - 4 = 6$ Ans. 6 pencils

(3) There were 8 sheets of paper. You used 5 sheets of paper for coloring. How many sheets of paper are left?
$8 - 5 = 3$ Ans. 3 sheets

(4) There are 7 eggs. You eat 3 eggs. How many eggs are left?
$7 - 3 = 4$ Ans. 4 eggs

(5) There were 9 sparrows on the roof. 2 sparrows flew away. How many were left?
$9 - 2 = 7$ Ans. 7 sparrows

① Read the passage. Then complete the exercises below.

In the spring, Rabbit and Pig plant a garden. First, Rabbit picks a spot in the yard. Then Pig rakes the soil and drops the seeds. Then Rabbit hops around to push the seeds down. Finally, Pig gets the hose from the house. He waters the garden.

(1) Circle the correct answers to each question.
ⓐ Who picks a spot?   ⓑ Who rakes the soil?
ⓒ Who hops around?   ⓓ Who waters the garden?

(2) Choose a word from the box to answer the questions.
last house spring yard
ⓐ When do Rabbit and Pig plant a garden?
They plant a garden in the spring.
ⓑ Where do Rabbit and Pig plant a garden?
They plant a garden in the yard.
ⓒ When does Pig water the garden?
Pig waters the garden last.
ⓓ Where does Pig get the hose?
Pig gets the hose from the house.

DAY 29, pages 57 & 58

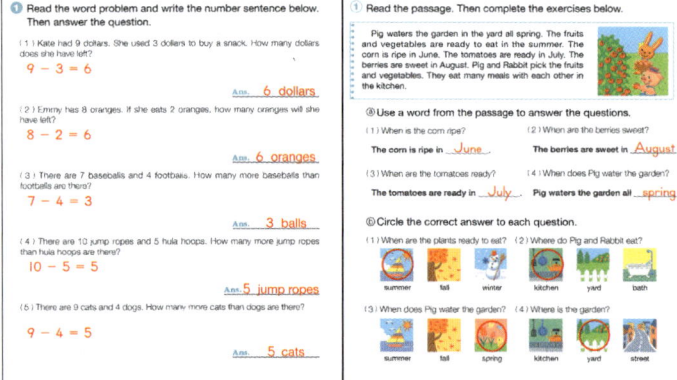

① Read the word problem and write the number sentence below. Then answer the question.

(1) Kate has 9 dollars. She used 3 dollars to buy a snack. How many dollars does she have left?
$9 - 3 = 6$ Ans. 6 dollars

(2) Emmy has 8 oranges. If she eats 2 oranges, how many oranges will she have left?
$8 - 2 = 6$ Ans. 6 oranges

(3) There are 7 baseballs and 4 footballs. How many more baseballs than footballs are there?
$7 - 4 = 3$ Ans. 3 balls

(4) There are 10 jump ropes and 5 hula hoops. How many more jump ropes than hula hoops are there?
$10 - 5 = 5$ Ans. 5 jump ropes

(5) There are 9 cats and 4 dogs. How many more cats than dogs are there?
$9 - 4 = 5$ Ans. 5 cats

① Read the passage. Then complete the exercises below.

Pig waters the garden in the yard all spring. The fruits and vegetables are ready to eat in the summer. The corn is ripe in June. The tomatoes are ready in July. The berries are sweet in August. Pig and Rabbit pick the fruits and vegetables. They eat many meals with each other in the kitchen.

ⓐ Use a word from the passage to answer the questions.
(1) When is the corn ripe?
The corn is ripe in June.
(2) When are the berries sweet?
The berries are sweet in August.
(3) When are the tomatoes ready?
The tomatoes are ready in July.
(4) When does Pig water the garden?
Pig waters the garden all spring.

ⓑ Circle the correct answer to each question.
(1) When are the plants ready to eat?
summer fall winter
(2) Where do Pig and Rabbit eat?
kitchen yard bath
(3) When does Pig water the garden?
summer fall spring
(4) Where is the garden?
kitchen yard street

DAY 30, pages 59 & 60

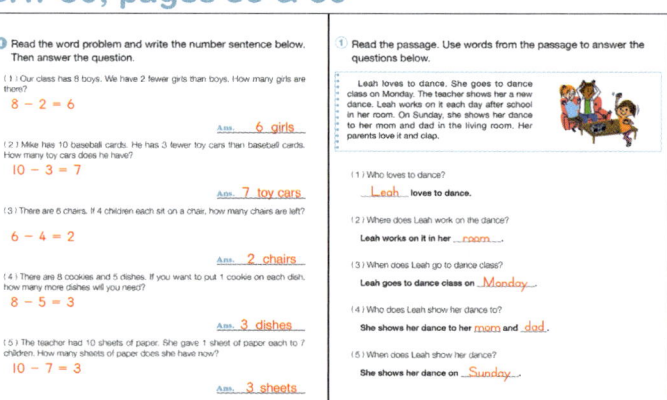

① Read the word problem and write the number sentence below. Then answer the question.

(1) Our class has 8 boys. We have 2 fewer girls than boys. How many girls are there?
$8 - 2 = 6$ Ans. 6 girls

(2) Mike has 10 baseball cards. He has 3 fewer toy cars than baseball cards. How many toy cars does he have?
$10 - 3 = 7$ Ans. 7 toy cars

(3) There are 6 chairs. If 4 children each sit on a chair, how many chairs are left?
$6 - 4 = 2$ Ans. 2 chairs

(4) There are 8 cookies and 5 dishes. If you want to put 1 cookie on each dish, how many more dishes will you need?
$8 - 5 = 3$ Ans. 3 dishes

(5) The teacher had 10 sheets of paper. She gave 1 sheet of paper each to 7 children. How many sheets of paper does she have now?
$10 - 7 = 3$ Ans. 3 sheets

① Read the passage. Use words from the passage to answer the questions below.

Leah loves to dance. She goes to dance class on Monday. The teacher shows her a new dance. Leah works on it each day after school in her room. On Sunday, she shows her dance to her mom and dad in the living room. Her parents love it and clap.

(1) Who loves to dance?
Leah loves to dance.
(2) Where does Leah work on the dance?
Leah works on it in her room.
(3) When does Leah go to dance class?
Leah goes to dance class on Monday.
(4) Who does Leah show her dance to?
She shows her dance to her mom and dad.
(5) When does Leah show her dance?
She shows her dance on Sunday.

DAY 31, pages 61 & 62

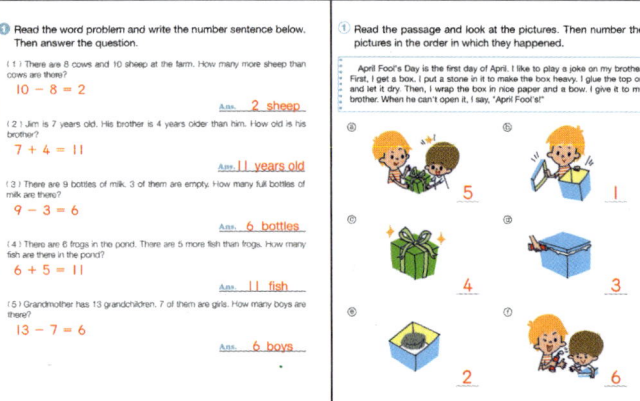

① Read the word problem and write the number sentence below. Then answer the question.

(1) There are 8 cows and 10 sheep at the farm. How many more sheep than cows are there?
$10 - 8 = 2$ Ans. 2 sheep

(2) Jim is 7 years old. His brother is 4 years older than him. How old is his brother?
$7 + 4 = 11$ Ans. 11 years old

(3) There are 9 bottles of milk. 3 of them are empty. How many full bottles of milk are there?
$9 - 3 = 6$ Ans. 6 bottles

(4) There are 6 frogs in the pond. There are 5 more fish than frogs. How many fish are there in the pond?
$6 + 5 = 11$ Ans. 11 fish

(5) Grandmother has 13 grandchildren. 7 of them are girls. How many boys are there?
$13 - 7 = 6$ Ans. 6 boys

① Read the passage and look at the pictures. Then number the pictures in the order in which they happened.

April Fool's Day is the first day of April. I like to play a joke on my brother. First, I put a box. I put a stone in it to make the box heavy. I glue the top on and let it dry. Then, I wrap the box in nice paper and a bow. I give it to my brother. When he can't open it, I say, "April Fool's!"

5   1
4   3
2   6

DAY 32, pages 63 & 64

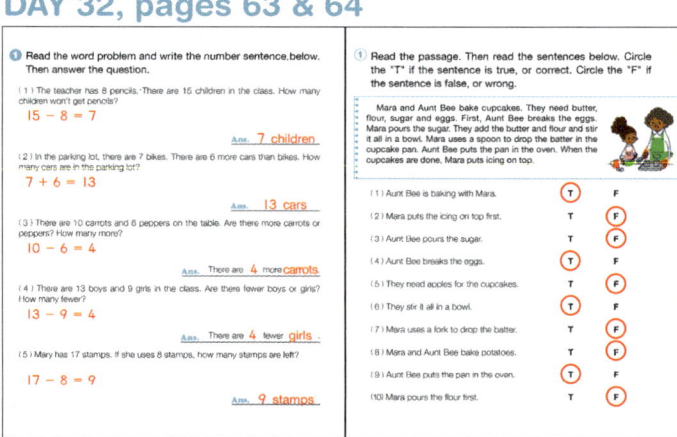

① Read the word problem and write the number sentence below. Then answer the question.

(1) The teacher has 8 pencils. There are 15 children in the class. How many children won't get pencils?
$15 - 8 = 7$ Ans. 7 children

(2) In the parking lot, there are 7 bikes. There are 6 more cars than bikes. How many cars are in the parking lot?
$7 + 6 = 13$ Ans. 13 cars

(3) There are 10 carrots and 6 peppers on the table. Are there more carrots or peppers? How many more?
$10 - 6 = 4$ Ans. There are 4 more carrots

(4) There are 13 boys and 9 girls in the class. Are there fewer boys or girls? How many fewer?
$13 - 9 = 4$ Ans. There are 4 fewer girls

(5) Mary has 17 stamps. If she uses 8 stamps, how many stamps are left?
$17 - 8 = 9$ Ans. 9 stamps

① Read the passage. Then read the sentences below. Circle the "T" if the sentence is true, or correct. Circle the "F" if the sentence is false, or wrong.

Mara and Aunt Bee bake cupcakes. They need butter, flour, sugar and eggs. First, Aunt Bee breaks the eggs. Mara pours the sugar. They add the butter and flour and stir it all in a bowl. Mara uses a spoon to drop the batter in the cupcake pan. Aunt Bee puts the pan in the oven. When the cupcakes are done, Mara puts icing on top.

(1) Aunt Bee is baking with Mara.   **T**   F
(2) Mara puts the icing on top first.   T   **F**
(3) Aunt Bee pours the sugar.   T   **F**
(4) Aunt Bee breaks the eggs.   **T**   F
(5) They need apples for the cupcakes.   T   **F**
(6) They stir it all in a bowl.   **T**   F
(7) Mara uses a fork to drop the batter.   T   **F**
(8) Mara and Aunt Bee bake potatoes.   T   **F**
(9) Aunt Bee puts the pan in the oven.   **T**   F
(10) Mara pours the flour first.   T   **F**

DAY 33, pages 65 & 66

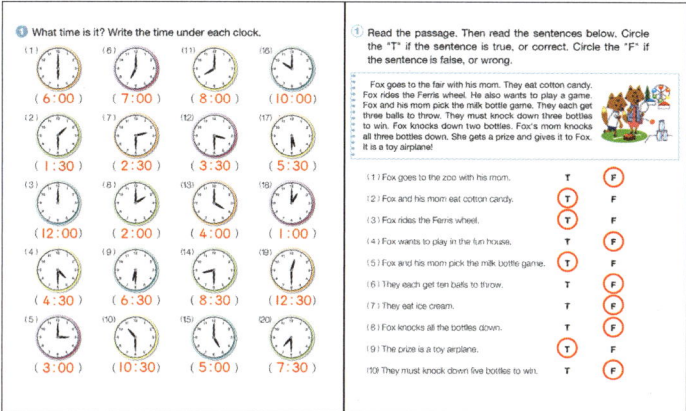

DAY 34, pages 67 & 68

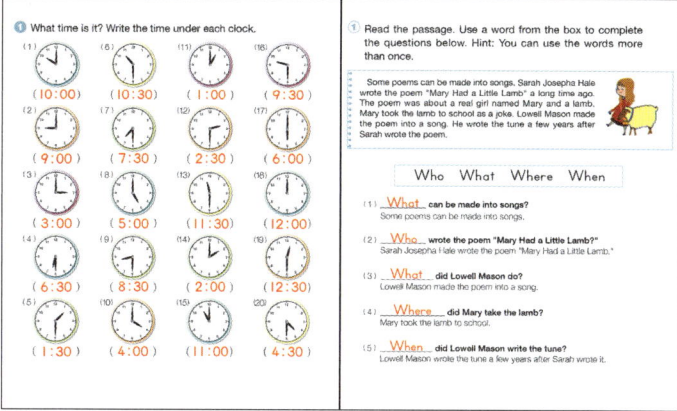

DAY 35, pages 69 & 70

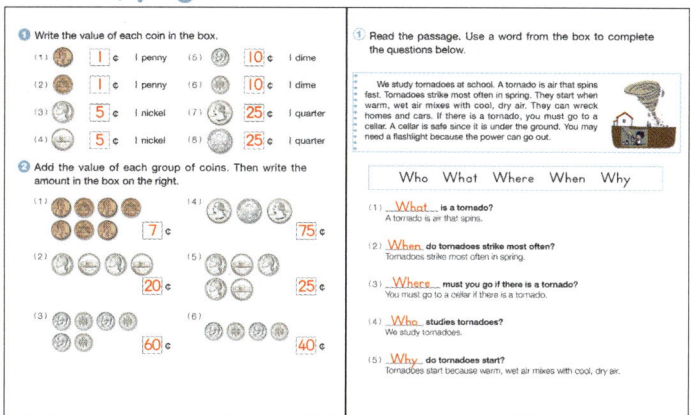

DAY 36, pages 71 & 72

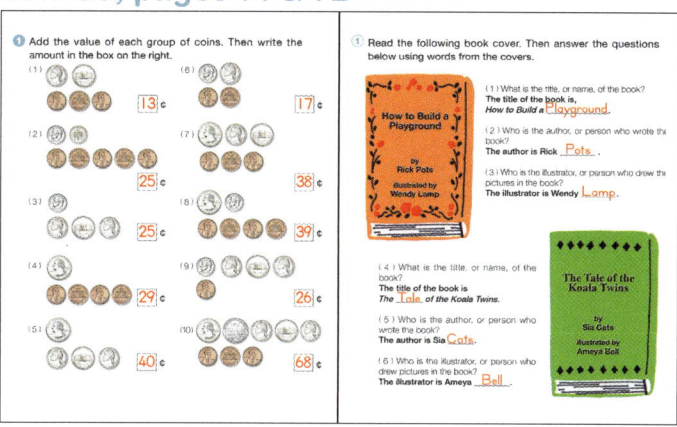

DAY 37, pages 73 & 74

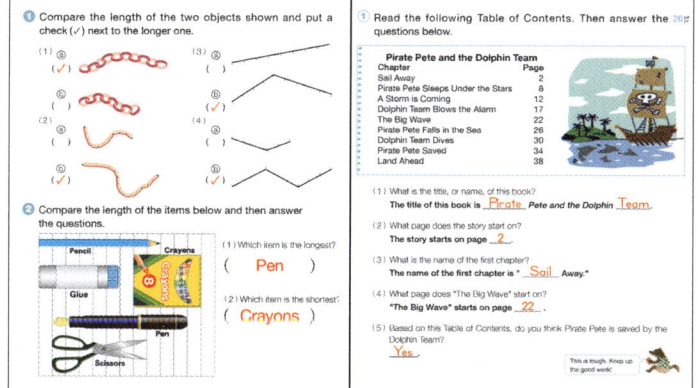

DAY 38, pages 75 & 76

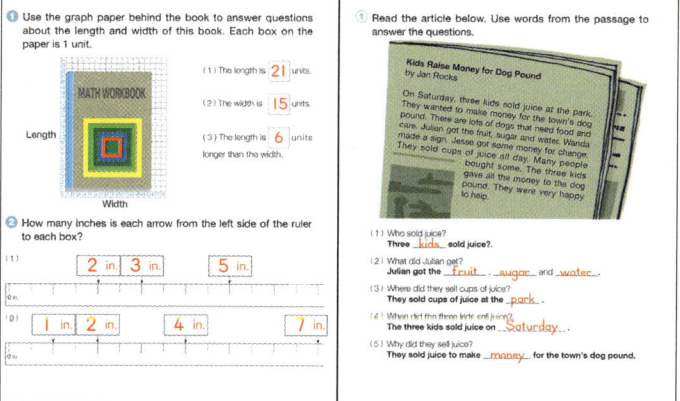

DAY 39, pages 77 & 78

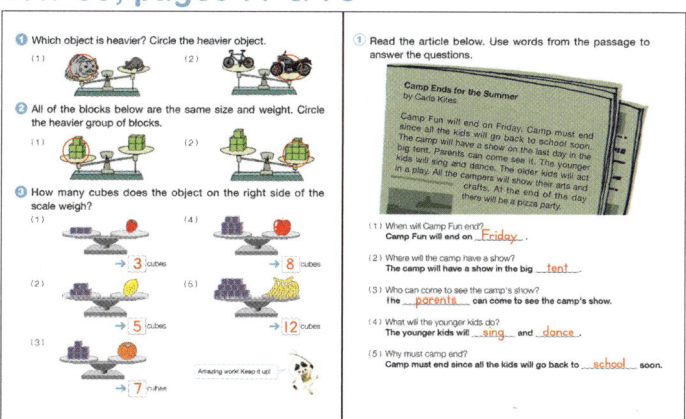

DAY 40, pages 79 & 80

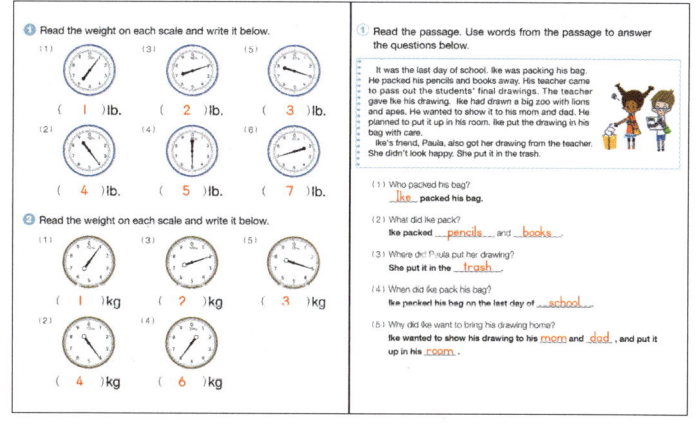

DAY 41, pages 81 & 82

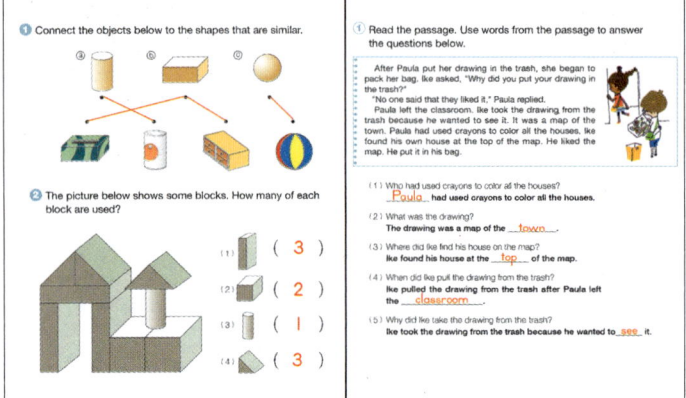

① Connect the objects below to the shapes that are similar.

② The picture below shows some blocks. How many of each block are used?

(1) (3)
(2) (2)
(3) (1)
(4) (3)

① Read the passage. Use words from the passage to answer the questions below.

After Paula put her drawing in the trash, she began to pack her bag. Ike asked, "Why did you put your drawing in the trash?"
"No one said that they liked it," Paula replied.
Paula left the classroom. Ike took the drawing from the trash because he wanted to see it. It was a map of the town. Paula had used crayons to color all the houses. Ike found his own house at the top of the map. He liked the map. He put it in his bag.

(1) Who had used crayons to color all the houses?
__Paula__ had used crayons to color all the houses.
(2) What was the drawing?
The drawing was a map of the __town__.
(3) Where did Ike find his house on the map?
Ike found his house at the __top__ of the map.
(4) When did Ike pull the drawing from the trash?
Ike pulled the drawing from the trash after Paula left the __classroom__.
(5) Why did Ike take the drawing from the trash?
Ike took the drawing from the trash because he wanted to __see__ it.

DAY 42, pages 83 & 84

① How many triangles ▲ are used to create the figures below?
(1) (3) (2) (4) (3) (2)

② You used three triangles (▲) to create the figures below. Draw two lines to separate the three triangles in each shape.
(1) (2) (3) (4)

③ Draw lines to separate the same-size triangles in each shape. How many same-size triangles were used to create the figures below?
(1) (6) triangles (2) (5) triangles

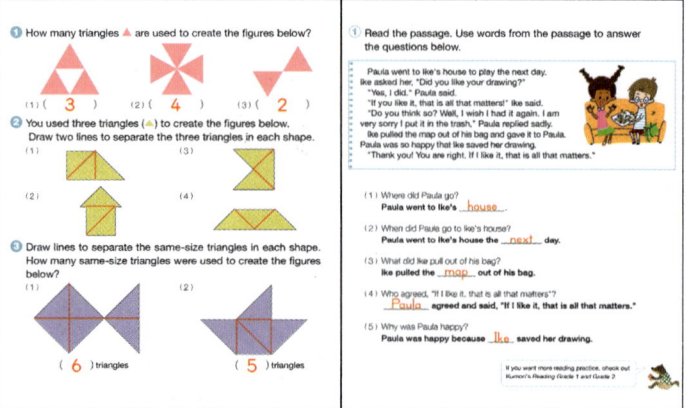

① Read the passage. Use words from the passage to answer the questions below.

Paula went to Ike's house to play the next day. Ike asked her, "Did you like your drawing?"
"Yes, I did," Paula said.
"Do you think so? Well, I wish I had it again. I am very sorry I put it in the trash," Paula replied sadly. Ike pulled the map out of his bag and gave it to Paula. Paula was so happy that Ike saved her drawing.
"Thank you! You are right. If I like it, that is all that matters."

(1) Where did Paula go?
Paula went to Ike's __house__.
(2) When did Paula go to Ike's house?
Paula went to Ike's house the __next__ day.
(3) What did Ike pull out of his bag?
Ike pulled the __map__ out of his bag.
(4) Who agreed, "If I like it, that is all that matters"?
__Paula__ agreed and said, "If I like it, that is all that matters."
(5) Why was Paula happy?
Paula was happy because __Ike__ saved her drawing.

If you want more reading practice, check out Kumon's Reading Grade 1 and Grade 2

DAY 43, pages 85 & 86

① Add or subtract.
(1) 9 + 2 = 11
(2) 11 − 5 = 6
(3) 15 + 3 = 18
(4) 13 − 2 = 11
(5) 8 + 6 = 14
(6) 14 − 6 = 8
(7) 7 + 4 = 11
(8) 10 − 4 = 6
(9) 9 + 8 = 17
(10) 13 − 8 = 5
(11) 14 + 5 = 19
(12) 15 − 5 = 10

② Read the word problem and write the number sentence below. Then answer the question.

(1) There are 5 candies on the dish and 7 candies in the box. How many candies are there in all?
5 + 7 = 12   Ans. 12 candies
(2) There are 9 chairs. If 6 children sit on chairs, how many are left?
9 − 6 = 3   Ans. 3 chairs
(3) You have 8 stickers. If I give you 3 stickers, how many stickers will you have in all?
8 + 3 = 11   Ans. 11 stickers
(4) There are 10 apples and 6 oranges. How many more apples than oranges are there?
10 − 6 = 4   Ans. 4 apples

① Write the correct vowels to finish each word below. Use the pictures as hints.
(1) man (2) jet (3) lip (4) job (5) hug
(6) fan (7) peg (8) big (9) log (10) cup

② Complete each sentence using a word from the box below.
buffalo wish tent wand lock stamp lunch bunch

(1) You can __wish__ upon a star.
(2) Don't __lock__ yourself out of the house.
(3) Put a __stamp__ on the envelope.
(4) Our __lunch__ was yummy.
(5) We put the __tent__ up.
(6) I grabbed a __bunch__ of crayons.
(7) They watched the __buffalo__ in the field.
(8) The fairy held a __wand__.

DAY 44, pages 87 & 88

① Add or subtract.
(1) 12 − 5 = 7
(2) 11 − 7 = 4
(3) 13 + 2 = 15
(4) 10 − 4 = 6
(5) 11 − 3 = 8
(6) 10 − 2 = 8
(7) 10 + 5 = 15
(8) 15 + 4 = 19
(9) 14 + 1 = 15
(10) 12 − 8 = 4
(11) 11 − 6 = 5
(12) 13 − 3 = 10

② Read the word problem and write the number sentence below. Then answer the question.

(1) Julie is 7 years old. Her brother is 6 years older than her. How old is her brother?
7 + 6 = 13   Ans. 13 years old
(2) There are 14 boys and 9 girls in a class. Are there fewer boys or girls? How many fewer?
14 − 9 = 5   Ans. There are 5 fewer girls

③ What time is it? Write the time under each clock.
(1) (12:00) (2) (1:30) (3) (9:00) (4) (3:30)

① Fill in the missing vowels. Use the pictures as clues.
(1) play (2) cube
(3) kite (4) read
(5) rain (6) meet
(7) toad (8) glue
(9) boat (10) bite

② Finish each sentence with a word from the box that describes the picture.
long heavy clean old fast

(1) My hair is not short.
It is __long__.
(2) The car is not new.
It is __old__.
(3) The dish is not dirty.
It is __clean__.
(4) This book is not light.
It is __heavy__.
(5) A lion is not slow.
It is __fast__.

DAY 45, pages 89 & 90

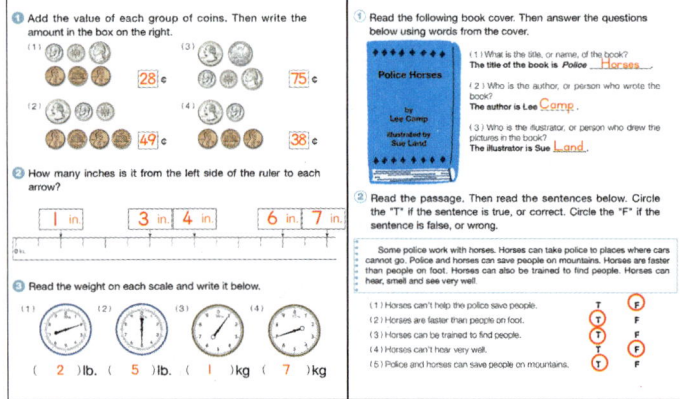

① Add the value of each group of coins. Then write the amount in the box on the right.
(1) 28 ¢   (3) 75 ¢
(2) 49 ¢   (4) 38 ¢

② How many inches is it from the left side of the ruler to each arrow?
1 in.   3 in.   4 in.   6 in.   7 in.

③ Read the weight on each scale and write it below.
(1) (2) lb. (2) (5) lb. (3) (1) kg (4) (7) kg

① Read the following book cover. Then answer the questions below using words from the cover.

Police Horses
by Lee Comp
illustrated by Sue Land

(1) What is the title, or name, of the book?
The title of the book is Police __Horses__.
(2) Who is the author, or person who wrote the book?
The author is Lee __Comp__.
(3) Who is the illustrator, or person who drew the pictures in the book?
The illustrator is Sue __Land__.

② Read the passage. Then read the sentences below. Circle the "T" if the sentence is true, or correct. Circle the "F" if the sentence is false, or wrong.

Some police work with horses. Horses can take police to places where cars cannot go. Police and horses can save people on mountains. Horses are faster than people on foot. Horses can also be trained to find people. Horses can hear, smell and see very well.

(1) Horses can't help the police save people.   T (F)
(2) Horses are faster than people on foot.   (T) F
(3) Horses can be trained to find people.   (T) F
(4) Horses can't hear very well.   T (F)
(5) Police and horses can save people on mountains.   (T) F